Post-Harvest Value Chain Management Including Logistics

CONTENT VIBES
PUBLISHERS ▪ PRINTERS ▪ AGGREGATORS

About the Author

Dr. Sarvesh Kumar was born at Etawah, Uttar Pradesh (India). He did his B.Sc. (Ag) Hons & M.Sc. (Ag) from CSAUA&T Kanpur, UP during 2001-2007. He received his Ph.D. in Extension Education from BHU, Varanasi (UP). He faced ARS twice in 2009, and 2010 and qualified ASRB (NET) respectively. Presently he is working as an Associate Professor, the Department of Extension Education, Institute of Agricultural Sciences, BHU, Varanasi (UP), India. He has worked as a Scientist (Agricultural Extension & Rural Sociology) for 12+ years (2011 to 2024) at Jawaharlal Nehru Krishi Vishwavidyalay, Jabalpur (MP) which has merit of the best Agricultural University in India (2018). He is the recipient of a Book Prize & Certificate from CSAU&T, Kanpur for being a topper of the Department of Agricultural Extension, M.Sc. (Ag) batch 2007. He has also received the Mandi Parishad Merit Scholarship during his B.Sc. (Ag) Hon's & M.Sc. (Ag) and the UGC Research Scholarship during his PhD at BHU, Varanasi (UP). He was designated as a member of the Anti-ragging Committee and elected representative of Research Scholars of the Institute of Agricultural Sciences, BHU, Varanasi (UP), India during 2007-2010. The excellent extension, teaching and research work done and many times was recognized by his host institution and district collector including an appreciation letter given by the Hon'ble Vice Chancellor, Agricultural minister, Minster to women and child developed Govt. of MP, and other national & International societies. His teaching research temperament may be witnessed by his research publication in nut shell as NAAS rated research papers 44, books 05, book chapters 12, extension bulletins-12, folder-40, articles English-20, articles hindi-131, ICT developed -03, varieties registration assistance under PPV&FRA 09, teaching-RAWE program & AIA-12 years, total award received-16, total appreciations & recognition-16, FPO developed -05, Common service centers developed-05, value chain development-10, HRD training received for 98 days, successful case & Impact registered by self-15, sponsored training given 239 days, In-service training-12, OFTs-12, FLDs-07 organized, training for farmers & farm women-538, training for rural youth-16, field days-48, Mela exhibitions-46, recognized fellowship received-03, invited lecturer given-554, paper presented national/international level-33.

Dr. Neeraj Singh is an alumnus of B.H.U. and School of Management Sciences, Varanasi, as he graduated from the Faculty of Commerce, B.H.U., in 2008 and PGDM in 2010. Dr. Neeraj obtained his PhD degree from the Institute of Management Studies, BHU in 2023. Dr. Singh has a rich +12 years of teaching and 2 years of corporate experience. Currently engaged as an Assistant Professor in MBA (Food and Agribusiness) formerly MABM, in the Department of Agricultural Economics, Institute of Agricultural Sciences, Banaras Hindu University. Dr. Neeraj has a keen interest in the fields of Human Resource Management, Communication, Agribusiness Environment and Policy, Strategic Management, Business Law and Ethics.

He has published more than fifteen research papers in international and national journals, 1 patent and 5 book chapters to his credit. He has attended 35+ conferences, and seminars and presented more than 29 papers at the event, and has also delivered lectures in winter schools organized by the department. He has also attended numerous FDP of one-week and three-week durations.

Post-Harvest Value Chain Management Including Logistics

Sarvesh Kumar
Associate Professor
Department of Extension Education
Institute of Agricultural Sciences
Banaras Hindu University
Varanasi-221005, UP

Neeraj Singh
Assistant Professor
Department of Agricultural Economics
Institute of Agricultural Sciences
Banaras Hindu University
Varanasi-221005, UP

CONTENT VIBES
PUBLISHERS • PRINTERS • AGGREGATORS

CV CONTENT VIBES
PUBLISHERS • PRINTERS • AGGREGATORS

1168, Sector 13, Urban Estate
Karnal-132 001, Haryana
Tel: 91-84470 75807, 18440 41168
Email: contentvibesppa@gmail.com
www.contentvibes.in

Print ISBN: 978-93-67550-93-9

ebooks ISBN: 978-93-67558-60-7

Preface

The agricultural sector plays a vital role in ensuring food security and economic stability, yet a significant portion of agricultural produce is lost due to inefficiencies in post-harvest management. Effective post-harvest value chain management, including logistics, is essential for minimizing losses, maintaining quality, and ensuring that food reaches consumers in optimal condition. This book explores the critical aspects of post-harvest handling, storage, transportation, processing, and distribution, emphasizing the importance of an integrated value chain approach.

Post-harvest management is a complex process that involves multiple stakeholders, including farmers, traders, processors, logistics providers, and retailers. The efficiency of this value chain determines not only the profitability of agricultural enterprises but also the affordability and availability of food for consumers. By implementing modern storage solutions, cold chain logistics, and digital traceability systems, India's agricultural supply chain can overcome traditional challenges such as food wastage, inadequate infrastructure, and market fluctuations.

This book delves into key topics such as packaging techniques, cold storage systems, transportation networks, supply chain financing, and regulatory frameworks governing post-harvest logistics. Additionally, it highlights the role of government initiatives, technological advancements, and public-private partnerships in strengthening India's agricultural value chain. Case studies and real-world examples are included to provide practical insights into successful post-harvest management strategies.

With globalization and evolving consumer demands, there is an urgent need to enhance the efficiency and sustainability of post-harvest practices. This book serves as a valuable resource for policymakers, academicians, agribusiness professionals, and students who seek a comprehensive understanding of post-harvest value chain management. By integrating best practices and innovative solutions, stakeholders can contribute to a more resilient and efficient agricultural supply chain, ensuring better returns for farmers and improved food security for the population.

I hope this book provides meaningful insights and fosters discussions on transforming post-harvest value chain management in India and beyond.

Authors

Contents

1

Understanding the Concept of Post-Harvest Value Chain

The Post-Harvest Value Chain encompasses the various steps agricultural products go through after being harvested until they reach the final consumer or market. This chain is important in influencing the product's quality, quantity, and market value. It involves important stages such as harvesting, handling, storage, processing, packaging, distribution, and retail. Each of these stages adds value, either by improving the product's quality, enhancing its shelf life, or converting it into a ready-to-sell finished product. Efficient management of the post-harvest value chain helps minimize losses, boosts food security, and increases profits for farmers and businesses. It also ensures that consumers receive high-quality, safe, and fresh food, making it an essential part of the agricultural sector.

DEFINNING THE POST-HARVEST SYSTEM

A post-harvest system refers to the process of managing agricultural products once they have been harvested. It involves activities such as cleaning, sorting, packing, and distributing the products to consumers. Post-harvest management focuses on the handling, storage, and transportation of agricultural commodities after harvest. For certain products like coffee and cocoa, additional post-harvest tasks, such as drying and fermenting, may also be necessary.

The post-harvest system should be viewed as the delivery process that moves crops from the point of harvest to their final consumption, aiming for minimal loss, maximum efficiency, and optimal returns for all parties involved. The term "system" refers to a dynamic, complex set of interconnected functions within a specific activity, while "chain" or "pipeline" highlights the sequential nature of operations, often overlooking their intricate interactions.

When considering the system or agro-food chain in its entirety, harvesting acts as the pivotal point between the pre-harvest phase (which corresponds to production) and the post-harvest phase (which extends from harvesting to consumption).

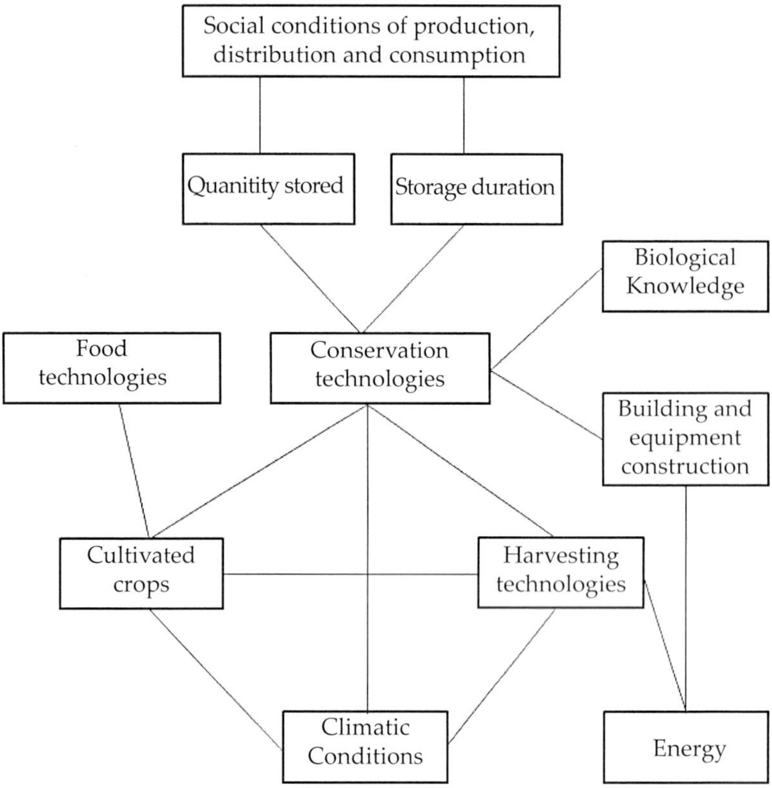

Figure: 1.1 Winnowing wheat, Afghanistan, FAO, 1994

The post-harvest system consists of a sequence of activities and operations that can be divided into two groups:

- **Technical Activities:** harvesting, field drying, threshing, cleaning, additional drying, storage, processing;
- **Economic Activities:** transporting, marketing, quality control, nutrition, extension, information and communication, administration and management.

KEY ELEMENTS OF THE POST-HARVEST SYSTEM

- **Harvesting:** The timing of harvest is based on the maturity level of the crops. For cereals and pulses, it is important to differentiate between the maturity of the stalks (straw), ears or seedpods, and the seeds, as this impacts subsequent steps, especially storage and preservation.
- **Pre-harvest drying:** Pre-harvest drying, especially for cereals and pulses, can improve preservation but also increases the risk of losses due to factors like pests (birds, rodents, insects), weather conditions, moulds, and theft. Harvesting before full maturity carries the risk of mould and seed decay.

- **Threshing:** Harvesting grains before they are sufficiently dry can result in incomplete threshing. If grain is damp and stored immediately after threshing (in granaries or bags), it becomes more prone to microorganism attack, reducing its preservation quality.
- **Post-Harvest Drying:** The duration required for proper drying of ears and grains depends on weather and atmospheric conditions. Drying structures like cribs or uncovered threshing floors expose the harvest to risks such as damage from livestock, birds, rodents, and other small animals. Moreover, these pests can leave droppings, which cause further losses. If grain is not properly dried, it can develop mould and rot in storage. Conversely, overly dry grain becomes brittle and may crack during threshing, hulling, or milling, leading to significant losses. Broken grains are more susceptible to pests like flour beetles and weevils. Over-drying also results in weight loss, reducing the monetary value at sale.
- **Transport:** Careful handling during transport is essential to prevent grain loss before reaching storage or threshing areas. The collection and initial transportation depend on where and how the harvest will be stored, particularly with regard to threshing.
- **Storage:** Effective storage requires proper facilities, hygiene, and monitoring. In sealed structures (granaries, warehouses, hermetic bins), controlling cleanliness, temperature, and humidity is crucial. Damage from pests (insects, rodents) and mould can degrade both the facility and the stored crop, leading to quality and quantity losses.
- **Processing:** Over-hulling or threshing can cause grain damage, particularly with rice, which can suffer cracks and lesions. This not only decreases the value of the grain but also makes it more susceptible to insect infestations like the rice moth *(Corcyra cephalonica)*.
- **Marketing:** Marketing plays a critical role in the post-harvest system and can take place at various stages in the agro-food chain, often during processing. It is closely linked to transport, which serves as a vital connection in the system.

POST-HARVEST LOSSES

"Losses are a measurable reduction in foodstuffs and may affect either quantity or quality". They arise from the fact that freshly harvested agricultural produce is a living thing that breathes and undergoes changes during post-harvest handling.

Loss should not be mistaken for damage, as damage is the visible indication of deterioration, such as chewed grains, and generally represents a partial issue. Damage limits the product's usability, while loss renders it entirely unusable.

Before delving into the various types of loss, it is important to establish some basic definitions.

Foodstuff refers to products, in this case, crops, that are edible by humans, particularly the parts that are fit for human consumption. In tropical regions, cereals and pulses account for about 75 percent of basic food, while the remaining vegetable-based foods are often sourced from roots and tubers, particularly cassava, yam, taro, plantain, potato, and sweet potato. In the food chain, food quantities are typically expressed by weight, though it is important to recognize that the organic structure and nutrients are just as significant.

Grains and seeds include cereals, pulses, and oilseeds cultivated across various climates for human consumption. The primary cereals are wheat, maize, rice, barley, sorghum, millet, oats, and rye; pulses encompass various types of peas, beans, broad beans, and lentils; and oilseeds include soybeans, groundnuts, sesame, rapeseed, and sunflower.

Post-harvest refers to the period after harvesting, which begins when the products are removed from the field at maturity and extends to the point of culinary preparation. Harvesting may be delayed for several months, especially for crops like maize and rice, in order to allow the straw and grain to fully dry. In such cases, the term "post-production" is sometimes used to highlight the connection between harvesting and post-harvest activities.

Food loss describes the total alteration or reduction in food quantity or quality, rendering it unfit for human consumption.

Variants of Losses

The first distinction in agro-food losses is between quantity and quality. Quantitative loss refers to a decrease in physical substance, such as a reduction in weight or volume, and can be measured and assessed. In contrast, qualitative loss pertains to the degradation of the food's value, particularly its nutritional and reproductive properties, which requires a different evaluation method.

It is important to clarify that losses caused during the production phase by various pests (such as insects, weeds, and diseases) will not be considered here, even though they significantly impact food preservation conditions and contribute to the extent and nature of post-harvest losses.

After discussing direct and indirect losses, we will explore weight loss, quality loss, food loss, seed viability loss, and commercial loss. Additionally, we must acknowledge the role of moisture content and the distinction between damage and loss.

Moisture Content

In bio-chemical terms, organic products consist of dry matter and water. Moisture content refers to the amount of free water in a product, usually expressed as a decimal or percentage. For example, cereals with a moisture content of 13% are considered to have good preservation potential. In agriculture, moisture content is usually given as a proportion of the moist product, indicating the weight of moisture relative to the total weight of both dry matter and moisture.

Damage

Damage represents a clear deterioration in a product, such as broken or damaged grains, which affects the quality more than the quantity and can ultimately lead to a complete loss. Both damage and loss should be measured in terms of weight and cost.

Direct and Indirect Losses

Direct losses occur when food is lost through leakage (such as spillage from bags) or consumption by pests (like insects, rodents, or birds), whereas indirect losses occur when a reduction in quality results in consumer rejection or refusal to purchase.

Weight Loss

Weight loss is easily observed and measured, but it doesn't always result in food loss. It may simply be due to a reduction in moisture content. For instance, moisture loss during drying is not considered a food loss. However, an abnormal increase in weight due to moisture absorption from rain can cause significant damage and result in a loss.

Weight loss can also happen from leakage during transportation, especially if sacks have holes or are poorly secured. Prolonged infestation and consumption by pests, or poor packaging, can also lead to weight loss. Pests' weight loss effects may not be immediately noticeable and could mislead an inexperienced buyer. A way to check for such losses is by comparing the weight of clean, healthy grain to that of a potentially infested sample, milling both, and comparing the resulting flour weights.

Quality Loss

Quality loss encompasses a wide range of factors, including external features like shape and size, as well as odor and taste. Cultural factors influencing food preferences and habits also play a role.

The cleanliness and overall health of a product are crucial for market acceptance and are typically defined as "sound, legal, and merchantable" products in commercial law. For instance, a trader may inspect a handful of grain for dust, which could indicate insect infestation. Similarly, a bad smell might suggest rodent contamination, which can be confirmed by the presence of droppings or fur. Other contaminants, such as bad grain, plant residues, soil, pebbles, or glass fragments, can reduce a product's value. Some contaminants, like pest excreta, oils, pesticides, or pathogenic organisms, pose a greater risk to health and safety.

Foreign bodies that distort the weight of a product also affect its quality, which in turn impacts its market value.

Food Loss

Food loss results from a decrease in both quantity and quality, ultimately making the product unfit for human consumption. Staple foods, which contain essential nutrients and vitamins, are especially susceptible. For example, grains are rich in protein and vitamins but are often targeted by rodents and insects.

The different parts of products, particularly their nutritive components, attract various pests. Weevils, for instance, feed on the carbohydrate-rich endosperm of seeds, while other pests target the vitamin-rich cereal covers. Vitamin levels are also affected by humidity during storage and fungal infections.

Seed Viability Loss

Seeds set aside for sowing must be carefully preserved to maintain their full germination potential. The grain heart, which is rich in protein, is often targeted by pests. Environmental factors, such as temperature, light, and humidity, also influence seed viability by causing excessive respiration, which diminishes the seed's productive potential.

Commercial Loss

Commercial loss refers to the financial impact of various types of losses, translating them into economic terms. While food prices are often determined by weight, several factors affect market value, particularly the qualitative aspects such as cleanliness and purity. Products with higher quality will be more sought after, especially when supplies are plentiful.

A producer's ability to capitalize on scarcity and raise the price of a commodity is a key aspect of good marketing. This requires proper organization and foresight, considering factors like adequate storage capacity and understanding long-term trends, such as socio-cultural changes, that may influence future demand.

While market-oriented skills are essential, the collective aspect of marketing, including training and preparation, also plays a key role in a producer's ability to thrive in the economy.

Unavoidable Unavoidable Losses and Compensation Strategies

Certain losses, referred to as "irreducible" losses, arise from respiration or mechanical rubbing of grains, and the breakage caused by specific machinery. Therefore, it is essential to remember that losses, whether pre-harvest or post-harvest (during production, distribution, storage, or marketing), cannot be completely eliminated. They must be compensated by increasing production. For example, compensating for a 20% loss requires a 25% increase in production, while a 40% loss requires a 66% increase, and a 60% loss demands a 150% increase in production.

Minimizing Post-Harvest Losses in Agricultural Value Chains

Post-harvest losses refer to the reduction in the quantity or quality of food products that occurs from the time of harvest until consumption. These losses happen during various stages of the agricultural value chain, including handling, storage, transportation, processing, and distribution. Contributing factors to post-harvest losses are diverse and include inadequate infrastructure, poor handling practices, limited access to markets, and environmental factors such as fluctuations in temperature and humidity.

According to recent estimates from the Associated Chambers of Commerce of India, India incurs a loss of approximately INR 926 billion (USD 14.33 billion) due to post-harvest losses.

Preventing Physical Damage During Harvesting and Handling

Physical damage to agricultural produce during harvesting and handling can occur due to improper techniques, equipment, or tools. Rough handling, mechanical injury, bruising, and crushing can cause deterioration in fruits, vegetables, grains, and other commodities. Harvesting at inappropriate maturity stages or under unfavorable weather conditions can further intensify this physical damage.

Limitations of Storage Facilities and Infrastructure

Insufficient storage facilities, such as warehouses, silos, and cold storage units, play a crucial role in post-harvest losses. Poor storage conditions, including lack of proper ventilation, temperature fluctuations, and high humidity, can promote spoilage, mold growth, and pest infestations. The absence of reliable electricity and refrigeration, particularly in remote or rural regions, exacerbates this issue.

Pests and Diseases

Pests and diseases are major threats to stored agricultural products, leading to spoilage and deterioration in quality. Insects, rodents, fungi, and bacteria can infest grains, fruits, and vegetables, causing contamination, mold growth, and decay. Without effective pest management strategies and proper hygiene practices, the losses caused by pests and diseases can be substantial.

Teansportational and Logistical Challenges

Inefficient transportation systems and logistics contribute to post-harvest losses by increasing transit times, exposing perishable goods to adverse conditions, and raising the chances of damage and spoilage. The lack of proper handling facilities, poor packaging, and rough handling during transit further worsen losses. Additionally, delays in transportation due to traffic, road conditions, or logistical challenges can compromise the quality and freshness of agricultural produce.

These causes highlight the complex nature of post-harvest losses, underlining the need for comprehensive solutions to address each contributing factor effectively.

Consequences Post-Harvest Losses

Economic Losses for Farmers and Stakeholders

Post-harvest losses cause significant economic setbacks for farmers, producers, and other stakeholders in the agricultural value chain. Reduced yields, lower market prices for damaged or spoiled produce, and increased production costs due to the wastage of inputs like seeds, fertilizers, and labor result in financial challenges. These losses not only affect farmers' livelihoods and incomes but also impact the profitability and sustainability of agricultural enterprises, impeding economic growth and development in rural areas.

Food Security Implications

Post-harvest losses have severe consequences for food security, especially in regions where food availability is already scarce. The loss of vital food items such as fruits, vegetables, grains, and pulses intensifies food shortages, malnutrition, and hunger. The reduced access to safe and nutritious food due to these losses undermines efforts to achieve food security and nutrition goals, perpetuating cycles of poverty and food insecurity in vulnerable populations.

Environmental Impacts

Post-harvest losses contribute to environmental degradation and resource wastage, increasing the ecological footprint of agriculture. The decomposition of spoiled produce releases greenhouse gases like methane (CH_4) and carbon dioxide (CO_2) into the atmosphere, driving climate change and global warming. Furthermore, the inefficient use of natural resources such as water, land, and energy in the production of wasted food places additional strain on ecosystems and exacerbates environmental issues.

Socio-Economic Consequences for Communities

Post-harvest losses have socio-economic effects on communities, leading to diminished livelihood opportunities, reduced income, and greater vulnerability to poverty and food insecurity. In rural areas where agriculture is the primary livelihood, decreased agricultural productivity hampers rural development and perpetuates poverty. Additionally, food loss and waste exacerbate inequalities, as marginalized communities often bear the brunt of food insecurity and nutritional deficiencies.

Approaches to Minimize Post-Harvest Losses

Enhancing Harvesting and Handling Practices

By adopting proper harvesting techniques and handling practices, farmers can significantly reduce physical damage to crops during the harvesting and post-harvest handling stages. Training programs and extension services can guide farmers on best practices, such as harvesting crops at the optimal maturity stage, using the right tools and equipment, and minimizing bruising and mechanical injury to produce during handling.

Upgrading Storage Facilities and Infrastructure

The lack of adequate storage facilities and infrastructure is a major factor contributing to post-harvest losses. Investing in modern storage solutions such as warehouses, cold storage units, silos, and drying facilities can help preserve the quality and extend the shelf life of harvested crops. Ensuring proper storage conditions—such as controlling temperature, humidity, ventilation, and pest management—is key to preventing spoilage and deterioration of stored produce.

Adopting Effective Pest and Disease Management Strategies

Pests and diseases represent significant threats to harvested crops during storage and transportation. To reduce losses caused by pests and diseases, integrated pest management (IPM) practices should be implemented. These include using biological control agents, crop rotation, and maintaining sanitation measures. Early detection and ongoing monitoring of pests and diseases, along with prompt interventions like chemical treatments or biopesticides, are key to prevent outbreaks and reduce damage.

Increasing Transportational and Logistical Efficiency

Efficient transportation and logistics are crucial for reducing transit losses and ensuring the timely delivery of produce to markets. Investments in transportation infrastructure—such as roads, railways, and cold chain logistics—can improve supply chain operations' efficiency and reliability. Adopting proper packaging materials and handling techniques during transit, including careful stacking and loading, helps reduce physical damage and spoilage. Additionally, leveraging technologies like GPS tracking, temperature monitoring, and real-time data analytics can increase visibility and traceability throughout the transportation process, helping to proactively manage logistical challenges and reduce losses.

Impact of Digitalization and Data-Driven Approaches

Digitalization and data-driven approaches are transforming post-harvest management by offering real-time monitoring, predictive analytics, and decision support systems that optimize supply chain operations. IoT (Internet of Things) devices, sensors, and smart monitoring systems provide continuous monitoring of essential factors like temperature, humidity, and gas concentrations in storage facilities and transport vehicles. By gathering and analyzing data on environmental conditions, product quality, and logistics performance, stakeholders can identify risks, spot anomalies, and take proactive actions to prevent losses. Additionally, digital platforms and mobile applications promote smooth communication, collaboration, and information exchange among farmers, suppliers, distributors, and retailers, streamlining supply chain processes and improving transparency.

Traceability solutions are important in post-harvest management, offering end-to-end visibility and accountability within the supply chain. Technologies such as blockchain, RFID (Radio Frequency Identification), and barcode scanning enable stakeholders to track the journey of produce from farm to table, capturing key data like origin, farming practices, handling methods, and quality characteristics. These traceability systems improve transparency, accountability, and authenticity, allowing for the quick identification and recall of contaminated or compromised products, thereby lowering food safety risks and building consumer trust. Furthermore, traceability data supports product differentiation, quality assurance, and compliance with regulatory standards, opening up opportunities for value-added marketing and premium pricing.

Streamlined Procurement

Streamlined procurement processes play a vital role in minimizing post-harvest losses by ensuring the efficient and timely sourcing of agricultural produce from farmers.

- **Reduced Harvest Delays:** Streamlined procurement processes allow buyers to communicate their demand and specifications to farmers ahead of time. This enables farmers to plan their harvest schedules accordingly, ensuring that produce is harvested at its optimal ripeness and maturity. This reduces the chances of overripening, spoilage, and quality deterioration in the field, helping to minimize post-harvest losses.

- **Optimized Supply Chain Logistics:** Streamlined procurement facilitates improved coordination between farmers, suppliers, and logistics providers. By accurately forecasting demand and aligning transportation schedules, buyers can ensure harvested produce is swiftly collected from farms and transported to processing or distribution centers. This reduces transit delays, minimizes handling and storage times, and prevents supply chain bottlenecks that can result in spoilage and losses.
- **Improved Quality Control:** Streamlined procurement processes often include quality assurance measures and standards for agricultural produce. Buyers can specify quality criteria, such as size, color, texture, and freshness, ensuring that only high-quality produce meeting these standards is purchased. This reduces the likelihood of substandard or damaged produce entering the supply chain, cutting down rejection rates and minimizing post-harvest losses.
- **Enhanced Traceability and Transparency:** Streamlined procurement processes utilize digital platforms and traceability solutions to monitor the movement of produce from farm to fork. By recording details such as harvest dates, farm locations, handling practices, and transportation routes, stakeholders can trace the origin of each batch and track its journey in real time. This transparency allows for the early detection of issues like delays, deviations from specifications, or quality concerns, enabling stakeholders to take proactive steps to prevent losses and ensure the integrity of the product.

2

Study of Existing Supply Chain of Different Commodities

The examination of existing supply chains for various commodities is essential for understanding the movement of goods, services, and information from the point of origin to the final consumer. Every commodity, whether raw materials like oil and metals or finished products like electronics and food, has its own distinct supply chain structure, which can differ in complexity based on factors such as geographic location, technological developments, regulatory frameworks, and market demand. Analyzing these supply chains helps pinpoint bottlenecks, inefficiencies, and areas for improvement, as well as identifying risks related to supply disruptions, price fluctuations, and environmental challenges. By examining current supply chains, businesses, policymakers, and researchers can develop strategies to optimize operations, reduce costs, enhance sustainability, and improve the overall resilience of supply systems in an increasingly interconnected global economy.

Agriculture is often not considered in the same way as other industries, as many people assume it simply involves farmers cultivating and selling crops, which are then used by various industries to produce their goods. However, agriculture encompasses much more, with many companies establishing their own farms. The food industry is experiencing rapid growth, projected to generate US $895.70 billion in revenue by 2024, with an annual growth rate of 7.67% (CAGR 2024-2029), and stakeholders are continuously innovating to increase production.

Even the agricultural sector has now implemented streamlined operations through supply chain management systems that efficiently manage the relationships between businesses involved in the production and distribution of products from farms to consumers. The main aim of streamlining these procedures is to fulfill consumer demands for quality, quantity, and price.

OPTIMZING SUPPLY CHAIN OPERATIONS

Supply chains primarily focus on the movement of goods and information between businesses involved in sourcing raw materials, converting them into finished products, and distributing them to end consumers. In the case of fresh agricultural products, the supply chain starts with farmers or planters who sell their harvest to primary processors. These processors then prepare the produce, package it, and ship it to wholesalers. Information drives supply chains and enables businesses to reduce costs, add value to products, optimize resource use, shorten time to market, and meet consumer demands. The success of a supply chain is determined by how well the operations across the chain are coordinated to create value for consumers while enhancing the profitability of each link. Essentially, supply chain operations form an interconnected process that delivers value to end consumers.

- **According to the Council of Supply Chain Management Professionals:** Supply chain management encompasses the planning and management of all activities involved in sourcing and procurement, conversion, and logistics. It also involves coordination and collaboration with channel partners, integrating supply and demand management both within and across companies.
- **According to Gartner:** Supply chain management refers to the processes of creating and fulfilling demand for goods and services. It involves a community of trading partners working together with the common goal of satisfying end customers.

Key Goals of Supply Chain Management

Supply chain management has five primary objectives that help organizations achieve their goals:

- **Add Value for Customers and Stakeholders:** Value refers to the worth of an item, product, or service. SCM aims to generate value for both supply chain stakeholders and end consumers.
- **Improve Customer Service:** Customer service measures a company's ability to address the needs, inquiries, and requests of customers and deliver products on time. Key aspects of customer service include:
    - Product availability for the customer.
    - Operational performance in fulfilling customer orders.
- Ensuring customer satisfaction.
- **Effectively Use System-Wide Resources:** Efficiency in supply chain operations means delivering the right product in the right quantity to the right customer at the right time. Resources include employees, raw materials, equipment, etc.
- **Efficiently Use System-Wide Resources:** Efficiency refers to the actual output compared to the expected standard output, typically expressed as a percentage. An efficient supply chain operates at minimal cost, within the time allowed, and with fewer resources while maintaining the desired output.
- **Leverage Partner Strengths:** A supply chain partnership is based on trust, shared risks, and rewards aimed at gaining a competitive advantage. A strong partnership provides:
    - Value addition, such as faster time to market.

- Improved market access, including new distribution channels.
- Financial and technological strength.
- Enhanced organizational skills.
- Reduced system costs and cycle times.

IMPORTANCE OF FRESH PRODUCE IN THE AGRICULTURAL SUPPLY CHAIN

The availability of fresh produce plays an important role for in-store businesses like supermarkets and convenience stores, as consumers tend to prioritize high-quality fruits and vegetables when selecting their primary retailers. However, emerging trends have raised concerns regarding the long-term sustainability of agricultural products. Issues such as unpredictable rainfall patterns, growing food demand, significant productivity gaps in farming compared to neighboring countries, and the challenge of balancing food prices with import dependencies to meet consumption needs have emphasized the urgent need for India to reevaluate its agricultural strategy moving forward.

Steps in Agriculture Supply Chain

The journey of a product from farm to market involves several critical phases within the agricultural supply chain:

Step 1: Production at the Farm

The process begins at the farm, where farmers carefully select the crops to grow or animals to raise, taking into account factors like soil health, weather patterns, and pest management. Modern agricultural practices also include the appropriate use of fertilizers and pesticides to ensure the crops and livestock develop optimally.

Step 2: Harvesting

Once crops have matured or livestock has reached the right age, harvesting takes place. This step may involve physical labor or specialized equipment, depending on the type of produce. Ensuring careful handling during this phase is vital for maintaining the quality and integrity of the product.

Step 3: Packaging and Grading

After harvesting, products are packed for protection during transportation. Packaging materials vary based on the type of crop, such as crates, cartons, or bags. Some crops are also graded by size, shape, and quality to ensure uniformity and streamline distribution, ensuring consumers receive consistent products.

Step 4: Transportation

Transportation plays a vital role in connecting farms to markets. Effective logistics systems move agricultural products from rural areas to urban centers. Long-distance transport often involves trucks, trains, ships, or planes, while local transportation uses smaller vehicles. Proper handling during this phase is essential to prevent damage and maintain freshness.

Step 5: Storage & Warehousing

Agricultural goods often require temporary storage before reaching their destination. Temperature-controlled warehouses and storage facilities help preserve the quality and freshness of perishable items. Maintaining optimal storage conditions is crucial to extending shelf life and preserving nutritional value.

Step 6: Processing and Value Addition

Some agricultural products undergo processing or value addition before they are sold. This can include turning raw goods into processed items, like canning fruits, milling wheat into flour, or producing oil from olives. Processing not only adds value but also extends the shelf life of products, offering opportunities for innovation within the agriculture sector.

Step 7: Distribution

Once ready for sale, agricultural products are distributed through a network of wholesalers, retailers, and other intermediaries who connect farmers with consumers. Efficient distribution systems ensure that products are quickly moved from storage to various market outlets, ensuring consistent availability to meet consumer demand.

Step 8: Retail and Consumer Access

The final step is when agricultural products are made available to consumers through retail outlets, farmers' markets, online platforms, or directly at farms. At this stage, consumers make informed purchasing decisions based on their preferences, dietary needs, and product information available through labels and certifications, which often detail the product's origin, manufacturing processes, and quality.

Key Elements of the Agriculture Supply Chain

The agriculture supply chain encompasses several key components, each playing a critical role in ensuring the movement of agricultural products from production points to consumers. These components include:

- **Manufacturing:** Involves the production of goods from raw agricultural materials.
- **Storage and Transportation:** Ensures the safe movement and preservation of agricultural products.
- **Processing and Packing:** Prepares agricultural products for sale by adding value and ensuring proper packaging.
- **Marketing:** Promotes and positions agricultural products in the market to attract consumers.
- **Selling:** The actual transaction of agricultural products to consumers.
- **Using:** The final consumption of agricultural products by individuals or industries.

Various stakeholders such as farmers, processors, transporters, and retailers facilitate the movement of agricultural products through each of these stages, ensuring they reach the end consumers.

Inventory Management

In agriculture, effective inventory management helps avoid both overstocking and stockouts, which can disrupt the flow of goods. This involves:

- Maintaining optimal levels of inventory to balance supply and demand.
- Ensuring product quality and quantity are properly quantified to minimize wastage.
- Managing storage costs efficiently while ensuring the right products are available at the right time to meet demand. Inventory management is essential in ensuring that the flow of products in the agricultural supply chain remains steady and cost-effective.

Fleet Management

Fleet management in agriculture focuses on the planning, coordination, and maintenance of vehicles used to transport agricultural goods. Proper fleet management helps ensure:

- Timely delivery of products to their destinations.
- Reduced transportation costs due to improved efficiency.
- Minimization of delays, leading to more reliable supply chain performance. Technology solutions such as GPS tracking and route optimization are integral to improving fleet performance, enabling real-time monitoring and enhancing the efficiency of logistics operations.

Supplier Management

Supplier management in agriculture is crucial for maintaining a consistent and quality supply of agricultural products. It includes:

- Identifying the right suppliers and establishing strong relationships.
- Negotiating terms of purchase agreements to ensure favorable conditions for both parties.
- Monitoring the quality and timeliness of deliveries to ensure that agricultural products meet standards. Contract management is key to ensuring that suppliers adhere to agreed quality and delivery standards, preventing supply disruptions and maintaining product consistency.

SIGNIFICANCE OF AGROFORESTRY IN COMMODITY SUPPLY NETWORKS

The integration of agroforestry into commodity supply chains is transforming the agricultural sector by promoting sustainability and fostering environmental stewardship. Agroforestry is a land management practice that combines trees and shrubs with crops or livestock, creating a harmonious system that boosts biodiversity, enhances soil fertility, and provides valuable ecosystem services.

The significance of agroforestry lies in its broad contributions to sustainable agriculture. It strengthens environmental resilience, helps mitigate the effects of climate change, and aids in biodiversity conservation. Moreover, agroforestry systems improve resource use efficiency, benefiting farmers economically through better yields and diversified income sources.

Commodity supply chains encompass the processes of producing, processing, and distributing goods, and understanding these chains is critical for ensuring sustainability. Agroforestry's integration enhances these supply chains by ensuring traceability, enforcing quality control, and promoting ethical practices, which together guarantee that commodities reach consumers in an environmentally responsible and socially fair manner. The inclusion of agroforestry in supply chains not only addresses ecological challenges but also strengthens the economic foundations of agricultural production.

Significance of Supply Chain Management (SCM) in Agriculture

- SCM enhances customer relationships and service quality.
- It boosts productivity and improves business operations.
- It helps reduce both direct and indirect costs.
- It minimizes product losses during transportation and storage.
- It provides valuable insights into product flow, market trends, and technological advancements.
- It enables better control over product safety and quality.
- It helps companies in minimizing waste and achieving greater efficiencies across the supply chain process.

The importance of supply chain management (SCM) in agriculture:

- SCM enhances customer relationships and improves service quality.
- It boosts productivity and streamlines business functions.
- It helps minimize both direct and indirect costs, making operations more cost-effective.
- SCM reduces product losses during transportation and storage by improving handling and logistics.
- It provides valuable insights into product flow, market conditions, and emerging technologies.
- SCM allows for better control over product safety and quality, ensuring consistency and compliance.
- It aids in minimizing waste and achieving greater efficiencies throughout the agricultural supply chain process.

Role of Agroforestry in Coffee, Rubber, and Cocoa Supply Chains

Agroforestry plays an important role in promoting sustainability within the supply chains of rubber, coffee, and cocoa cultivation. In rubber farming, integrating trees with rubber crops enhances soil health, prevents erosion, and provides shade, all of which promote healthier rubber tree growth. For coffee and cocoa, agroforestry creates an ideal microclimate by offering shade, which is vital for the delicate plants. Moreover, the diverse interactions between trees and crops foster increased biodiversity, effective pest control, and better pollination services. These benefits lead to improved yields and strengthen the resilience of these supply chains to environmental changes. By adopting agroforestry practices, the rubber, coffee, and cocoa industries take a holistic approach, balancing economic productivity with environmental and social sustainability. This transformative approach represents a significant step towards sustainable agriculture.

Climate Challenges and Vulnerability in the Cocoa Industry

Cocoa, the main ingredient in chocolate, is predominantly grown in tropical regions like West Africa, South America, and Southeast Asia. Cocoa trees require specific climatic conditions—high temperatures, abundant rainfall, and minimal temperature fluctuations—to thrive. However, climate change is disrupting these conditions, presenting significant risks to cocoa production.

- Shade trees help protect cocoa plants from excessive heat and reduce water loss through evapotranspiration, ensuring optimal growing conditions and improving the quality of cocoa beans.
- Cocoa agroforestry systems incorporate nitrogen-fixing trees that enhance soil fertility. This improves soil health, supports cocoa tree growth, and increases yields.
- By integrating other crops and forestry products, farmers can diversify their income sources, reducing their reliance on cocoa and enhancing economic stability.

Climate Sensitivity and its Impact on the Coffee Industry

Coffee is one of the world's most traded commodities and is highly sensitive to climate changes. Coffee plants need consistent rainfall and moderate temperatures to produce high-quality beans, and climate change is altering these conditions, potentially disrupting coffee supply chains.

- Shade trees help protect coffee plants from heat and reduce water evaporation, maintaining optimal growing conditions and improving bean quality.
- Tree roots enhance soil structure and fertility, supporting healthier coffee plants and boosting yields.
- Coffee agroforestry systems support diverse ecosystems, which naturally control pests and diseases, reducing the need for chemical interventions.

Impact of Climate on Rubber Industry

Rubber, primarily grown in tropical regions like Southeast Asia, is significantly impacted by climate change. Rubber trees need specific climatic conditions to produce latex, and deviations from these conditions can affect production.

- Rubber agroforestry systems improve soil fertility and reduce erosion, enhancing rubber tree growth.

- Shade trees provide optimal conditions for rubber tree growth and latex production.

Agroforestry enables the cultivation of additional crops or forestry products, diversifying income sources and reducing farmers' dependence on rubber alone.

Agroforestry in Commodity Supply Chains: Environmental, Economic, and Social Benefits

The integration of agroforestry into commodity supply chains provides important environmental advantages:

- **Carbon sequestration:** Trees in agroforestry systems capture and store carbon, mitigating climate change by offsetting greenhouse gas emissions associated with commodity production.
- **Biodiversity conservation:** Diverse tree-crop interactions create habitat niches, promoting biodiversity and fostering resilient ecosystems.

These environmental benefits not only contribute to sustainability but also enhance the resilience of commodity supply chains in the face of climate change.

Economic Benefits

Agroforestry systems bring several economic advantages:

- **Diversified Income Sources:** Agroforestry combines agricultural and forestry components, offering farmers multiple revenue streams. This helps mitigate financial risks related to market fluctuations or crop failures.
- **Resilience to Climate Change:** Trees provide a buffer against extreme weather, reduce soil erosion, and contribute to ecosystem resilience, enhancing the adaptive capacity of agricultural systems.

Agroforestry helps ensure long-term economic stability and sustainability for farmers facing climate challenges.

Social Benefits

Agroforestry also generates significant social benefits:

- **Community Empowerment:** Agroforestry practices often involve collective decision-making and resource management, empowering local communities.
- **Job Creation:** Agroforestry creates employment opportunities through tree planting, maintenance, and harvesting, fostering skill development within communities.

These social benefits contribute to rural development by strengthening community bonds, improving livelihoods, and promoting shared responsibility for environmental stewardship.

Overall, the integration of agroforestry into commodity supply chains provides a sustainable, resilient approach to agricultural production, benefiting the environment, economy, and society.

Key Considerations in Commodity Supply Chains

- **Rising Temperatures:** Higher temperatures stress cocoa, coffee, and rubber trees, reducing bean quality, yields, and latex production. For example, cocoa trees may produce lower-quality beans, while coffee plants may experience growth difficulties.

- **Irregular Rainfall:** Erratic rainfall patterns negatively affect crop growth. Cocoa plants become more susceptible to pests and diseases, while coffee plants face increased vulnerability to issues like coffee leaf rust. Similarly, rubber trees' growth and latex production are disrupted by inconsistent rainfall.
- **Deforestation:** Cocoa and rubber cultivation have historically contributed to deforestation, which worsens climate change. The expansion of rubber plantations and cocoa farming disrupts ecosystems, affecting biodiversity and productivity. Coffee cultivation, too, suffers from habitat loss, which harms the plants' growth and surrounding ecosystems.
- **Soil Depletion:** Intensive rubber cultivation leads to soil degradation, reducing fertility and affecting long-term crop productivity.

Future Trends and Opportunities in Agroforestry

- **Remote Sensing:** This technology is crucial for monitoring tree growth and health in agroforestry systems. It aids in early detection of issues, improving management practices.
- **Precision Agriculture:** Precision farming techniques optimize the use of resources like water, fertilizers, and pesticides, enhancing productivity and reducing waste in agroforestry systems.
- **Global Scale Adoption:** Agroforestry is well-positioned for expansion due to its adaptability and environmental benefits. It plays a key role in increasing supply chain resilience by addressing climate risks and ensuring resource availability.

Opportunities also lie in partnerships between technology firms and agribusinesses, which can streamline the implementation and monitoring of agroforestry projects, driving the global shift toward sustainable farming practices. As the world focuses on sustainability, the integration of agroforestry into global supply chains offers significant benefits in terms of increased yields, reduced environmental impact, and improved socio-economic outcomes, setting the stage for a more resilient and sustainable agricultural future.

Supply Chain Management in Dairy

The dairy supply chain encompasses the entire process of acquiring, transporting, transforming, storing, and delivering milk and dairy products to consumers. A critical aspect of this supply chain is maintaining the cold chain, which ensures the quality and safety of the products. This system helps in minimizing wastage and ensuring the timely delivery of nutritious, fresh dairy items that meet consumer demands.

Supply Chain Management in Poultry

In the poultry supply chain, key stages include breeding, hatching, raising, processing, packaging, and distributing poultry products. This chain focuses on the security of biological products, animal treatment, and ensuring the quality of the final products. Efficient supply chain management helps ensure timely deliveries, reduces losses, and meets the market demand for high-quality, fresh, and hygienically processed poultry items.

Supply Chain Management in Horticulture

The horticultural supply chain involves managing the production, harvesting, handling, storage, processing, packaging, and distribution of fruits, vegetables, and ornamental plants. The focus is on ensuring product quality and freshness, especially for food products on retail counters. Effective management also minimizes losses after harvest and guarantees that consumers receive fresh, high-quality horticultural produce.

OBSTACLES IN AGRICULTURAL SUPPLY CHAIN MANAGEMENT

Agriculture supply chains face numerous challenges, including extreme weather conditions, market access barriers, rising food demand, a growing preference for healthier foods, and labor issues. Addressing these problems requires innovation, strategic planning, and coordinated efforts from all relevant stakeholders.

Extreme Weather Conditions

Climate-related disruptions, such as droughts, floods, and storms, significantly impact crop production and transportation systems. These weather patterns can reduce agricultural yields, damage crops and finished products, and obstruct transportation and distribution networks. Addressing these challenges involves adopting resilient farming practices, risk assessment, mitigation strategies, and effective disaster response mechanisms.

Market Access and Trade Barriers

Barriers to market access, such as tariffs, trade restrictions, and inadequate infrastructure for the movement of agricultural products, limit farmers' opportunities and increase costs. Overcoming these challenges requires policy reforms, investment in infrastructure, and the promotion of international trade to expand market access.

Rising Demand Threatening Food Security

The growing global demand for food, driven by population growth and increased consumption, poses a threat to food security. To meet this demand, agricultural productivity needs to be enhanced, and food waste minimized, all while ensuring that sustainable farming practices are followed.

Increasing Demand for Healthier Food Choices

Consumers' increasing preference for healthy and organic foods, particularly locally produced items, puts additional pressure on agricultural supply chains. This demand requires maintaining high product quality, ensuring traceability, and meeting evolving consumer expectations. Supply chains must adapt, engage stakeholders, and collaborate to meet these changing demands.

Labor Issues

Labor-related challenges include shortages, low wages, and poor working conditions for farm workers, which can lead to inefficiencies, delays, and increased costs. Addressing these issues requires improving working conditions, ensuring fair compensation, and exploring mechanization and automation where possible to enhance productivity.

Key Obstacles in the Agricultural Supply Chain

The agricultural supply chain faces numerous obstacles that hinder its efficiency. The issues are compounded by a lack of traceability, excessive storage costs, and other challenges. Some of the significant hurdles include:

Poor Packaging Practices

A large number of goods are transported to markets or storage facilities without proper packaging or protection, which leads to substantial damage. Items are often tossed onto the ground for counting and price negotiations. Some degree of sorting and grading might take place at the location, but products are typically loaded back onto trucks for further transportation, adding to waste and losses.

Security and Safety Issues

Fresh produce logistics must comply with food safety regulations that govern transportation, storage, cleanliness, and warehousing. For instance, transportation vehicles must meet ISO standards and handle various cargo conditions. Non-compliance with these standards can result in financial and legal repercussions at different stages of the supply chain.

Excessive Preservation Needs

Fresh food supply chains require careful preservation to maintain quality. Goods need to be stored at correct temperatures, properly packaged, and handled promptly to prevent deterioration.

Challenges for Farmers

Farmers in India face multiple difficulties, with fair pricing being a significant issue. They struggle with soil and water health, limited knowledge about high-value crops, lack of exposure to modern farming techniques, poor market access, inefficient supply chains causing food waste, and heavy dependence on rainfall. To improve agricultural income, it's essential to streamline the entire value chain, from cultivation to the point where farmers can realize profits.

Liquidity Issues for Growers

The agricultural supply chain is often impacted by unethical trading practices, especially due to the imbalance between small-scale and large-scale farmers. Payment delays and fluctuating crop prices create financial challenges for farmers, leaving them unable to buy inputs for the next season. This leads to financial instability and forces farmers to take out loans with high-interest rates.

Inadequate Post-Harvest Infrastructure

The agricultural and horticultural supply chains suffer from major gaps in post-harvest infrastructure at the farm level. Activities like sorting, grading, washing, and packaging are often missing at the farm site. Without farm-level collection centers, additional losses occur, diminishing value in the supply chain and impacting farmers' profitability.

Maintaining Produce Shelf Life

Ensuring that fresh produce is stored in optimal moisture and temperature conditions is critical. Each fruit type has specific requirements for temperature and moisture to retain quality. However, with insufficient storage facilities, the shelf life of produce is shortened, causing quality deterioration. This results in significant losses for farmers, particularly in industrialized countries. Post-harvest handling processes often experience high wastage rates, sometimes reaching up to 30%.

Leveraging Digital Technology to Address Challenge

Given the increasing demand for food and the limited resources available, especially after the COVID-19 pandemic, it is crucial to develop a long-term vision for digital agriculture. Digital technologies offer tools to create a modernized supply chain that addresses existing challenges and opens new avenues of value. For example, India has one of the best soil health management systems globally, utilizing digital technologies throughout the production and post-production phases.

Merets of Modern Technology in Farming

Advanced Access to Inputs

By providing farmers with easier access to agricultural inputs near their locations, innovative tools and techniques can be adopted to enhance productivity and yield.

Financial Support

Government initiatives ensure that farmers facing financial difficulties can quickly access loans, helping to alleviate financial pressures.

Digital Agriculture

Specialized companies in digital agriculture offer advanced technological solutions that improve agricultural productivity and operational efficiency.

Enhanced Supply Chain Efficiency

Utilizing technology-driven, relationship-based platforms helps streamline the connection between agricultural products and potential buyers, thus enhancing the overall supply chain's efficiency.

Farming as a Service

This model, using subscription-based or pay-per-use methods, makes cutting-edge technology more affordable for farmers. It allows farmers to access valuable resources while converting fixed costs into variable costs, making modern agricultural solutions more accessible.

Enhancing Supply Chain Management for Perishable Agricultural Products

The significant losses of fruits and vegetables in the country are primarily due to inadequate storage and transportation infrastructure that fails to maintain controlled conditions. Recently, Supply Chain Management (SCM) has gained prominence with the advent of globalization. A supply chain consists of three or more interconnected organizations that are linked through the flow of products, services, finances, and information from the source to the customer. SCM advocates for a supply chain approach that involves actively managing the two-way movement and coordination of goods, services, information, and funds from raw materials to the final consumer. The evolving lifestyle and open economy have prompted manufacturers and suppliers to focus on producing and delivering high-quality products.

Several factors are driving the emphasis on supply chain management. First, the availability and cost-effectiveness of information resources between supply chain entities enable seamless connections that reduce time delays. Second, the growing competition in both domestic and international markets demands that organizations be swift, agile, and adaptable. Third, customer expectations are becoming increasingly stringent. To meet these demands, SCM systems must prioritize two main objectives: timeliness and quality.

The Supply Chain Umbrella

Supply chain management encompasses a wide range of activities beyond purchasing. Each of these diverse functions is part of an interconnected network that dictates how goods and information move efficiently across the supply chain. The key activities include:

- **Purchasing:** Purchasing is a core activity for most organizations, as it serves as the central point of focus within the supply chain.
- **Quality Control:** Nearly all organizations recognize the importance of supplier quality and the need to prevent quality issues rather than merely detecting them. Progressive organizations collaborate directly with suppliers to develop effective quality control procedures and processes.
- **Demand and Supply Planning:** Demand planning involves forecasting anticipated demand, making inventory adjustments, and addressing unfulfilled orders, including spare parts and after-market requirements. Supply planning takes this demand data and creates a supply, production, and logistics network to meet these needs.

- **Material or Inventory Control:** This group is usually responsible for managing the inventory levels of finished goods needed to support customer demand. Their focus is on the physical distribution side of the supply chain, including outbound or downstream logistics.
- **Order Processing:** This activity ensures that customers receive materials when and where needed. It acts as the link between the producer and the external customer.
- **Production Planning, Scheduling, and Control:** This function involves creating a time-phased production schedule, developing short-term production plans, and managing work-in-process production.
- **Warehousing/Distribution:** Especially important for companies that manufacture based on forecasts, warehousing and distribution focus on storing and moving goods in anticipation of future sales.
- **Customer Service:** Customer service includes a range of activities aimed at ensuring customer satisfaction with products or services, maintaining strong relationships with consumers.

TNAU & MSU COLLABORATION ON FRUIT & VEGETABLE SUPPLY CHAIN CHALLENGES

Tamil Nadu Agricultural University and Michigan State University Collaboration: Addressing Supply Chain Challenges in Fruits and Vegetables

Tamil Nadu Agricultural University (TNAU) and Michigan State University (MSU) have partnered to tackle the emerging supply chain challenges in the fruit and vegetable sector through the "Building University Capacity to Improve Fruit and Vegetable Supply Chain Development in India" project. This initiative, funded by USAID through the Association of Liaison Office (ALO), aims to enhance supply chain management (SCM) in India by:

- Developing faculty competencies
- Creating curriculum and learning opportunities
- Strengthening partnerships among stakeholders, including farmers

The partnership between TNAU and MSU has introduced unique aspects that elevate the project beyond typical value chain development. The focus not only addresses the value chain but also considers the bio-physical and socio-cultural environments in which these systems operate.

Success Stories

Federation Formed for the 'King of Fruits'

During a workshop at TNAU under the HED/ALO project, an action plan was developed after a SWOC analysis involving various mango stakeholders in Tamil Nadu. The consensus was that the supply chain could be made more effective with the formation of farmers' forums. Consequently, the "Tamil Nadu Mango Growers' Federation" (TAMAFED) was established to address issues related to mango production, fostering teamwork among farmers to tackle challenges in supply chain management.

Online Mango Trading - 'One India-One Market Concept'

Farmers in Krishnagiri now benefit from fair prices for their mangoes, thanks to the online spot trading introduced by Safal National Exchange (SNX) for Tothapuri mangoes. The initiative has enabled farmers to sell their produce globally, ensuring fair prices. In 2021, around 2,000 tonnes of mangoes were sold at premium prices.

Tamil Nadu Growers Achieve EurepGAP Certification

Progressive mango growers from Tamil Nadu, such as Mr. Paul Ponnen and Mr. Sethupathy, achieved EurepGAP certification for their gardens, enabling them to export fresh mangoes to the Gulf and Europe. This certification has encouraged other growers to follow suit and increase export volumes, aiming for 300 tonnes for future shipments.

Farmer-Owned Post-Harvest Complex for Bananas

Banana stakeholders, including growers and traders, recognized the need for better handling infrastructure. In response, the Tamil Nadu Banana Growers' Federation and the Theni Banana Growers and Traders Association requested the government to build a ripening and storage facility. The Agriculture Minister approved the creation of a 450-tonne capacity facility, worth Rs. 75 million ($1.87 million), to improve quality, extend shelf life, and broaden distribution. This model has inspired similar initiatives in Coimbatore and Erode.

Farmer-Owned Producer Company

The establishment of the 'Dharmapuri Precision Farmers Agro Services Ltd.'—a company owned by 180 farmers—marks a significant development in commercial empowerment. The farmers, now shareholders, have transformed into the CEOs of the company. The firm is planning to establish a wholesale mega farmers' market to create direct market linkages for their produce.

Establishing the Directorate of Agri-Business Development at TNAU

The success of the SWOC analysis and stakeholder engagement led to the establishment of the Directorate of Agri-Business Development at TNAU. This new directorate supports the needs of farmers and other partners in the supply chain, signifying a policy-level transformation within the university.

Developing Strong Market Linkages

Several interactive meetings were held to connect growers with major buyers like Reliance, Maagritaa, Safal, Field Fresh, and Dole, helping establish direct market linkages and opening new sales channels for farmers.

Accomplishments

On October 6, 2006, Ms. Radha Singh, IAS, the Agriculture Secretary of the Government of India, attended an interactive meeting with the office bearers of the Tamil Nadu Mango and Banana Federations, as well as participants of the "Tamil Nadu Precision Farming Project." This meeting, held alongside Mr. Surjit K. Choudary, Agriculture Secretary of Tamil Nadu, and Dr. C. Ramasamy, Vice Chancellor of TNAU, resulted in the submission of several key demands to the Government of India. Ms. Singh agreed to consider these requests, which included:

- A 50% subsidy for drip irrigation
- Subsidies for liquid fertilizers used in drip irrigation
- Waiving of taxes (cess) on the transport of perishable goods
- A separate color plate for vehicles carrying perishables
- Special interest rates for loans from nationalized banks for agriculture/ horticulture

Linkages

An interactive meeting was held on December 5-6, 2006, with officials from Field Fresh Foods Pvt. Ltd., DOLE Food Company, and the Tamil Nadu Banana Growers Association. The Dole representatives, including Mr. Nicomedes "Nikko" S. Ylanan, Manager of Quality Assurance, Mr. Mayo Edwin C. Javier, Mango Business Manager, and Mr. Augustine, visited potential banana fields owned by federation members. Dole expressed interest in assisting with the cultivation of 2,000 acres of bananas in Tamil Nadu.

Capacity Building

A two-day seminar titled "Supply Chain Management and Food Safety Issues" was held at HC & RI, Periyakulam, on September 10-11, 2007. The seminar aimed to disseminate knowledge about the project's findings and SCM issues related to fruits and vegetables to TNAU scientists. Thirty university scientists attended, with topics covering GAP certification, the benefits of EurepGAP certification, and experiences in mango cultivation. The seminar also included a visit to "Crystal Farms," a EurepGAP certified garden in Palani. This initiative had a significant impact on TNAU faculty, leading to the introduction of a "Supply Chain Management" course for B.Tech (Horticulture) undergraduates.

Future Prospects and Trends

With ongoing support from TNAU, the federations continue to advance the development of a modern supply chain for fruits and vegetables in Tamil Nadu. Special emphasis is placed on including small farmers in this process and forming additional partnerships with national and international markets. The federations are also focusing on becoming more sustainable by generating resources through membership fees and conducting workshops, fairs, and training programs. These efforts aim to enhance the quality, value, and returns of the farmers' produce. The federations also receive support from development funds provided by state and central governments, banks, private companies, cooperatives, national and international NGOs, the World Bank, and other development organizations.

3

Case Study and Analysis of Value Chain of Food Grains

The value chain of food grains encompasses a comprises of interconnected activities that enhance the raw food grains as they progress from farm production to consumption. Each stage in this chain contributes to the quality, availability, and cost of the final product.

A food value chain represents a network of businesses collaborating to cultivate, process, distribute, and sell food. It operates as a business model with the objective of boosting financial returns while also promoting social and environmental values.

DEFINING THE FOOD VALUE CHAIN

Food value chains are designed to enhance competitive advantage by fostering collaboration among various stakeholders in the process. A food value chain connects producers, processors, marketers, food service providers, retailers, and supporting organizations. It can be understood as a strategic partnership between interdependent businesses, coming together to gradually create value for the final consumer, which in turn results in a collective competitive advantage.

The food value chain (FVC) brings together farmers, processors, distributors, retailers, and consumers in a cooperative effort. With increasing demand for transparency regarding the entire process of bringing food from farms to tables—including production, processing, storage, transportation, and marketing—a food value chain utilizes affordable communication technologies to maintain constant and transparent communication among all partners within the chain.

The food value chain plays a significant role in global greenhouse gas emissions, contributing to about one-third of the emissions from agri-food systems. This sector involves all activities beyond the farm, such as:

- Transportation
- Processing

- Packaging
- Storage and distribution
- Retail
- Food service
- Household consumption
- Waste disposal

Key Facts about Agri-food Systems

- Agri-food systems account for 35% of global emissions.
- Emissions from the food value chain are increasing at a faster rate than farm-level emissions.
- Major sources of emissions include food waste (7.9%), household consumption (7.3%), and retail (4.2%).

As the global population grows and consumption patterns change, addressing emissions within the food value chain becomes essential for mitigating climate change. This is particularly critical in high- and middle-income countries, where these emissions are rising rapidly.

By focusing on energy efficiency, waste reduction, and sustainable practices throughout the food value chain, significant progress can be made in reducing overall agri-food system emissions, thereby supporting global climate objectives.

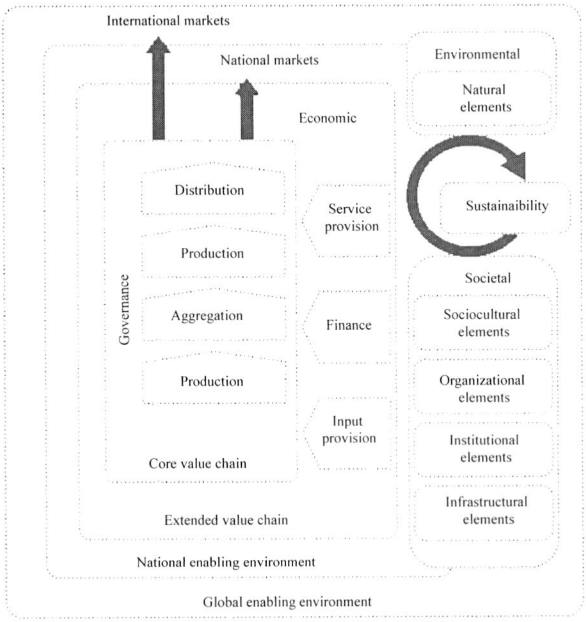

Figure: 3.1 Sustainable food value chain framework

According to the United Nations Environment Programme's (UNEP) Food Waste Index Report 2021, approximately 1 billion tons of food are wasted globally every year. This means that a third of all food produced worldwide is lost or wasted. Food supply chains are intricate and often opaque, involving multiple stakeholders such as farmers, government bodies, retailers, and logistics providers. These global disruptions highlight the urgent need for more sustainable and resilient food chains. In light of this, the traditional food value chain requires a comprehensive overhaul, which should be supported by the deployment of new technologies, the adoption of sustainable practices, and innovative approaches to minimize waste. Collaborative efforts will be essential to drive holistic growth. Key actions such as streamlining processes, improving cold chain infrastructure, and adhering to regulatory standards will help redefine the food supply chain and make it capable of meeting global market demands.

A usual value chain for the food processing industry involves several key stages that work together to produce and deliver food products to consumers. These stages include:

- **Farming:** The initial stage where crops are cultivated and livestock is raised to supply raw materials for the food processing industry.
- **Ingredients Procurement:** The process of sourcing raw materials such as fruits, vegetables, grains, and meats from suppliers to use in food production.
- **Multiple Levels of Food Processing:** This stage involves transforming raw materials into finished food products, through methods like canning, freezing, or packaging.
- **Quality Control and Food Safety:** Ensuring that food products meet established quality and safety standards. This includes activities like laboratory testing, inspections, and adherence to safety regulations.
- **Marketing and Distribution:** Promoting food products to customers and ensuring they reach the market through various distribution channels, such as supermarkets, restaurants, and online retailers.
- **Customer Service:** Addressing customer inquiries and providing support, which may involve managing product returns, handling complaints, and ensuring customer satisfaction.

By conducting a value chain analysis within the food processing industry, businesses can pinpoint areas where efficiency can be improved, costs can be reduced, and the overall customer experience can be enhanced. This comprehensive evaluation can ultimately lead to more streamlined operations and better profitability.

Understanding Value Chain Approach

The growing debate around the sustainability of our food system and its ability to address future challenges has sparked significant public interest in various activities within the food system. These activities occur within what are known as value chains. Food value chains encompass all processes, from the production of agricultural raw materials and their inputs to transportation, processing for retail, and finally, the delivery of products to consumers. Food chain management is also of considerable economic importance. In the European Union, around 280,000 companies are engaged in the food industry, with a combined sales volume of approximately 840 billion euros and providing about 4 million jobs.

Food Chain Management has gained heightened social relevance, particularly due to food scandals and the growing consumer demand for transparency and sustainability. Simultaneously, the food industry is navigating a more complex environment, characterized by increasing competition, legal regulations, and adherence to specific standards.

For society, the sustainability of these value chain activities has become a prominent topic of discussion. Concerns are raised regarding their environmental impact, potential health effects, and whether those working within the sector are receiving fair compensation. These issues gain further importance considering that value chains are becoming increasingly globalized. Ultimately, value chains represent a growing division of labor within society, and managing these chains effectively must be addressed both in practical applications and academic discussions.

The food supply chain in India is a complex network of participants and processes at different stages that work together to ensure that food products reach consumers. Both the formal and informal sectors play key roles in this system. key components that form the food supply chain in India are:

- **Consumer Demand:** Production is influenced by consumer demand, which is often overlooked but is crucial for planning. In India, consumers are categorized into four groups based on their food preferences:
  - **Vegans:** Individuals who consume only plant-based foods.
  - **Vegetarians:** Those who consume plant-based food along with milk and milk products.
  - **Eggetarians:** Vegetarians who eat eggs but avoid meat.
  - **Meat eaters:** Those who consume both vegetarian and non-vegetarian foods.
- **Food Production:** The food supply chain in India begins with agriculture. Farmers, often informal and unorganized, are the backbone of food production. Key products include:
  - **Crops:** Grown primarily for vegans and vegetarians.
  - **Livestock (Cows):** Raised for milk production for vegetarian consumers.
  - **Poultry Farming:** Focused on egg production for eggetarians and meat production for meat eaters.
  - **Livestock (Cattle):** Raised for meat production, with cows typically avoided for meat due to religious beliefs.
  - **Seafood:** Sourced through mariculture, particularly in coastal regions.
- **Processing:** After production, food goes through various processing stages before it is ready for consumption or distribution.
- **Distribution:** This stage involves various service providers and intermediaries who help deliver food from producers to consumers:

- **Transporters:** Logistics play a crucial role in India due to its vast and diverse landmass, facilitating the movement of food products.
- **Wholesalers:** They purchase unprocessed food from farmers and processed food from factories, acting as intermediaries and selling goods to retailers.
- **Retailers:** The final point of contact for consumers, where food products are sold to the public.

- **Consumption:** Final consumers purchase food from retailers and consume it.

In essence, the food supply chain in India is a collaborative system of individuals, processes, and intermediaries working together to ensure that food moves efficiently from farms to consumers.

Variants of Value Chains

A closer examination of the fresh food supply in industrialized countries reveals that there are numerous potential combinations leading to different types of value chains. In nearly every country, typical local value chains still exist, such as direct sales from agricultural holdings. This indicates that production and consumption remain largely local or regional, although some production inputs or resources are often sourced from other areas. In these local value chains, production, consumption, and waste disposal occur within the same region. Additional activities, such as craft businesses or service companies, also contribute to the functioning of these value chain processes.

Global value chains, on the other hand, are typically linked to large food manufacturers. Here, production and consumption are distinctly separated in terms of location, and to bridge the spatial and temporal gap, various actors are involved to maintain product quality and freshness. Recently, "coordinated value chains," which have emerged due to the rising significance of highly organized food retailing, have become a focal point of discussion. It is widely recognized that food retail chains play a leading role in contemporary value chains. Acting as consumer representatives, they set the quality standards and delivery requirements. However, because of the intense competition within the food retail industry, they also place a strong emphasis on efficiency, striving to implement this focus in the coordination of the value chain.

A Systemic Perspective

Value chains can be represented by a simplified model), which typically includes suppliers, producers, processors, or wholesalers, especially in the case of fresh products, who then resell to the food retail industry. From there, the product reaches the consumers. In the fresh product sector, food retailers play a significant role, particularly in Northern Europe, where they supply over 80% of the population's fresh produce. As a result, retail chains are dominant players in this system.

However, solely observing the value chain members does not provide a comprehensive understanding of food system outcomes. It overlooks other contributors such as service companies, the finance sector, and the increasingly vital consultancy sector, all of which support the functioning of value chains. Additionally, the logistics sector faces major challenges due to rising globalization. In many industrialized countries, the existing infrastructure is often taken for granted, potentially underestimating its role in value chain development. This infrastructure includes research and education systems, information and communication technology, transport networks (such as roads and railways), cooling facilities throughout the value chain, and control institutions that ensure compliance with product and process standards. The importance of infrastructure is particularly evident in developing countries, where it often presents a significant barrier to the growth and development of value chains.

Importance of Value Chain Coordination

If we assume that coordination within value chains happens solely through market transactions in ideal markets, there would be little need to focus on coordination challenges. However, issues related to quality and product safety demonstrate that market-based coordination alone is insufficient to prevent opportunistic behavior and improper product handling. To address this, both public and private efforts are often necessary to ensure product quality and food safety.

For value chains to be organized efficiently and cost-effectively, it is crucial to reduce transaction costs. This can be achieved, for example, by establishing commercial categories (such as grades). Direct communication and information exchange are also vital, particularly in local food systems. Emerging technologies, like radio frequency identification (RFID), are expected to play an important role in the future.

The ongoing debate surrounding climate change and the emissions generated by food systems has further highlighted the externalities associated with supply chains. Market coordination is often insufficient, and there are significant information gaps at every stage of the value chain. Reducing these external effects remains a major challenge for future value chain management. From a global perspective, it is crucial to address whether food systems can meet the growing world population's demand for healthy food, as this is directly connected to concerns about environmental pollution.

In global value chains, coordination is also viewed through the lens of competitiveness. In some markets, where there is an oversupply of products, the question arises whether it is possible to offer consumers benefits such as lower prices or differentiated products and services to stand out from competing value chains.

Governance Approaches for Value Chain Coordination

To succeed in the global agricultural and food market, it is important to have a deep understanding of the food itself, its production processes, marketing strategies, and consumer behavior. The complexity of food chains—spanning from producers to end consumers and involving multiple stakeholders with distinct requirements—necessitates a comprehensive approach to management.

This raises the question of how value chains can be effectively governed. Specifically, it is crucial to determine how various interdependencies can be managed. This challenge is often discussed in terms of governance or governance mechanisms within value chains. The aim is to analyze the "rules of the game" in the system and, from a practical standpoint, define these rules.

In markets that are not highly differentiated, market coordination remains vital, and direct communication continues to play a significant role, especially in more localized value chains. The rise of private labels in the food industry has also led to the growing importance of retail contracts between partners. This trend is particularly relevant in food production and processing, where large companies often source products on a contract basis. In these situations, more general forms of coordination are carried out through grades and standards. Product standardization, in the form of commercial grades, must be addressed, as well as the standardization of processes, such as the adoption of quality management systems by producers. Moreover, global and public regulations, also contribute to coordination. Overall, the coordination of value chains today depends on a dense network of diverse arrangements. Relevant mechanisms are situated between public and private institutions and always require thorough documentation on the production side, significantly affecting the structure and transformation of the agricultural sector.

Looking at the focus of food value chain coordination in recent years, several phases can be identified. In periods of intense competition, cost reduction and productivity were central concerns. Economies of scale, specialization, and increased productivity played major roles, supported by horizontal cooperation (marketing cooperatives). Regarding logistics, centralization and standardization gained importance, while vertical cooperation along the value chain intensified. Efficient process design became increasingly significant for corporate management.

As concentration within both production and retail value chains increased, these chains became less transparent and more anonymous. This shift became especially evident in the 1980s, when food scandals began to occur more frequently. While cost and productivity concerns remained, the importance of building consumer trust grew. Issues such as transparency, documentation, and traceability emerged as key concerns. These problems were addressed through certification systems and the development of process standards. Communication with consumers also became a major issue.

However, heightened competition necessitated greater product differentiation to cater to various consumer segments. Producer brands and private labels from food retailers gained more attention, not just in agriculture but across the entire food industry. To protect investments in these brands, companies established comprehensive contracts with suppliers. This process was supported by a significant increase in investment in advertising and other communication efforts.

Today, societal demands on the food system are increasingly prominent in public discussions, with a strong focus on the environmental challenges and climate change associated with food production. New actors, such as non-governmental fair trade organizations and environmental groups, have emerged within the food system. Companies are now expected to act responsibly and engage in dialogue with society. Trusted and transparent communication remains essential in this context.

It is also important to critically evaluate whether the focus on reducing costs and boosting production has compromised consumer interests and societal needs.

Segments of Agricultural Commodity Trade

Condiments and Sauces

In 2021, India's imports of condiments, sauces, jams, and jellies totaled $29.8 million, with nearly $6.8 million sourced from the United States. Other major suppliers included Thailand, China, Malaysia, the UAE, and several European nations. The demand for imported condiments and sauces is growing in India, as both consumers and restaurants experiment with foreign products, driving growth in the retail and restaurant sectors.

Cotton

India is one of the largest producers and exporters of cotton globally. However, it continues to import extra-long staple and high-quality long staple cotton (28-34 mm) and occasionally imports medium staple cotton when international prices are favorable. On April 13, 2022, India's government removed the 10 percent import duty on cotton until September 30, 2022, to support the textile industry amidst record-high domestic fiber prices. The United States has been a leading cotton supplier to India, with U.S. cotton exports reaching $214 million in 2021, a 31 percent increase from the previous year. U.S. cotton has been favored for its quality and consistency, especially in the medium to long staple categories. The U.S. has also been the leading supplier of Pima cotton since 2015. However, competition from Brazil, Egypt, and Australia is growing, due to factors like freight advantages and quicker delivery times.

Wines and Spirits

India's alcoholic beverage sector, valued at $35 billion, is the world's third largest. Despite high import tariffs, state-specific regulations, and occasional marketing limitations, the sector continues to expand. In 2021, imports of wines and distilled spirits surged to $319 million, marking a 40 percent year-on-year increase. U.S. exports of American wines and distilled spirits to India saw a remarkable 114 percent rise in 2021, reaching $7 million.

Craft Beer and Beer Ingredients

The Indian craft beer market is poised for significant growth, though it remains a niche segment. With over 200 microbreweries and brewpubs in operation, up from 45 in 2016, there is increasing demand for ingredients like malt, hops, and yeast, as well as imported beers. The microbrewery sector, though small, represents one percent of the total beer market. A few breweries are expanding their production lines to include bottled and canned craft beers, while consumer acceptance of craft beer grows across India in states where alcohol consumption is permitted.

Fresh Fruits

India offers market access for most fresh fruits, and demand for imported fresh produce is rising due to a growing segment of consumers who seek high-quality standards and year-round availability. In 2021, imports of U.S. fresh fruits, including apples, pears, cherries, and blueberries, were valued at $24.2 million, with apples leading at $22.5 million. However, this was a decline from 2020 due to a tariff increase on apples from 50 percent to 70 percent in response to U.S. steel and aluminum tariffs and pandemic-related trade disruptions. Despite this, imports of fresh fruits are expected to grow, with new products entering the Indian market. U.S. processed fruit exports also saw significant growth, including dried cranberries, prunes, jams, and jellies, which saw a 34 percent increase in 2021.

Hides and Skins

India's imports of hides and skins rose for the first time in six years, reaching $24.8 million in 2021, a 26 percent increase from the previous year. U.S. exports of hides and skins to India also increased by 58 percent to $853,000. Raw hides enter India with zero tariffs, while tanned leather imports are subject to a 15 percent duty. The raw material is primarily used by India's local leather industry, though some is re-exported.

Pulses

India is the largest global producer, consumer, and importer of pulses, with imports reaching $2.1 billion in 2021. In 2016, imports peaked at 6.99 million tons, valued at $3.9 billion, mostly from Canada, Burma, Australia, Russia, China, and the United States. However, the Indian government has imposed several measures to restrict pulse imports, including raising import duties from zero to 30-60 percent and imposing quantitative restrictions on key pulses. In April 2022, the government lifted restrictions on certain pulses to tackle food inflation, though imports of other pulses remain restricted. Historically, the U.S. has been a significant supplier of higher-priced pulses like green and yellow peas, lentils, and chickpeas, but due to trade restrictions, imports have dropped sharply from 369,000 metric tons in 2014 to just 12,500 metric tons in 2021.

Snack Foods

As consumer lifestyles evolve and disposable incomes increase, demand for imported snack foods in India continues to rise, despite competition from local brands. The market for sweets and snacks is expanding rapidly, aided by rising incomes and holiday gifting traditions. The pandemic accelerated the adoption of digital marketing and delivery platforms, fostering partnerships with organized retailers. Manufacturers are procuring high-value specialty ingredients to meet the growing demand for healthier snack options. In fiscal year 2019-2020, the industry reached sales of over $13 billion, and though the pandemic led to a significant reduction in sales, it is expected to recover and reach $8.56 billion in fiscal year 2022.

Tree Nuts

Tree nuts are a significant U.S. agricultural export to India. In 2021, imports of U.S. tree nuts to India surpassed $889 million, with almonds (primarily in-shell) being the dominant export, valued at $750 million. India also imports considerable amounts of walnuts, pistachios, and hazelnuts, primarily from the United States, Iran, Afghanistan, and the Middle East. Almonds, in particular, are gaining popularity among India's growing middle-income population due to their health benefits.

PHASES OF FOOD PROCESSING

For instance, by optimizing the ingredient sourcing process, a food processing company can lower its costs and improve the quality of its raw materials. Additionally, investing in advanced food safety technologies can help bolster its reputation for providing safe and high-quality products.

By prioritizing customer needs, leveraging data and technology, collaborating with suppliers, investing in research and development, implementing a sustainability strategy, and fostering innovation, companies in the food processing industry can enhance their competitiveness and profitability.

- **Center the Customer:** The customer should be the primary focus when analyzing the value chain. Understanding their needs and preferences is essential for determining the value created at each stage of the chain. For instance, if customers prioritize convenience, a food processing company might focus on creating products that are easy to prepare and cook quickly.
- **Prioritize Research and Development:** Research and development are essential for the growth of the food processing industry, as they lead to the creation of new and innovative products. For example, investing in R&D might help develop new preservation techniques or identify alternative, more sustainable raw material sources.
- **Partner with Suppliers:** Strong collaboration with suppliers is key to the success of the food processing industry. By forming close relationships, companies can ensure access to high-quality raw materials, lower costs, and enhance the sustainability of their operations.
- **Leverage Data and Technology:** Data and technology are vital for the food processing industry. By gathering and analyzing data on key operations, companies can pinpoint areas for improvement and make informed choices. For example, data analytics can help optimize supply chain management, reduce waste and costs, and enhance the quality of raw materials.
- **Create a Sustainability Plan:** Sustainability is becoming increasingly important to consumers, and food processing companies must craft strategies to address this. This could involve minimizing waste and emissions, using renewable energy, and improving supply chain sustainability.
- **Adopt Innovation:** The food processing industry is continuously changing, and companies must embrace innovation to stay competitive. This could involve investing in new technologies, developing new products, or discovering innovative ways to reach customers.

Significance of Each Activity in Value Chain

Evaluating the significance of each activity in the value chain involves assessing its contribution to customer value, cost, competitive advantage, supply chain considerations, market demand, and adherence to regulatory and industry standards. This analysis provides valuable insights that help businesses prioritize activities and make well-informed decisions to enhance their operations and strengthen their competitive position in the market.

Below are the key factors to consider when evaluating the importance of each activity in the food processing industry:

- **Customer Value:** The most critical factor in determining the importance of an activity is its ability to contribute to customer value. Activities that directly impact the quality, safety, and attractiveness of the final food product hold greater significance. For instance, quality control and food safety measures are crucial in ensuring that the product meets high standards, which is essential for maintaining customer trust and satisfaction.
- **Cost:** The cost associated with each activity is another important consideration. Activities that incur higher costs may still be deemed more significant if they play a key role in enhancing the quality and safety of the food product. Balancing cost with value creation is essential for efficient operations.
- **Competitive Advantage:** The ability of each activity to provide a competitive edge should also be taken into account. For example, a food processing company may achieve a market advantage if it has a strong reputation for food safety and quality, or if it possesses unique processing technologies that differentiate it from competitors.
- **Supply Chain Considerations:** The impact of each activity on the overall efficiency and sustainability of the supply chain is another important factor. Activities that help reduce waste, lower emissions, or enhance the sustainability of raw materials are considered more vital for improving both operational efficiency and environmental responsibility.
- **Market Demand:** The importance of an activity may also be shaped by shifts in market demand. For example, if consumer demand for organic and sustainable food products rises, activities related to the sourcing and procurement of organic or sustainable raw materials may gain greater significance.
- **Regulation and Industry Standards:** Compliance with regulatory and industry standards is a key consideration in evaluating activity importance. For example, food safety regulations may make quality control and food safety activities more critical, while sustainability standards could elevate the importance of developing a comprehensive sustainability strategy.

Examining Competitive Edge

Differentiation: It is essential to evaluate how each activity contributes to distinguishing a company's food products from its competitors. Activities that significantly enhance the quality, safety, and appeal of the final food product can give the company a competitive advantage. For example, a company with stringent quality control and food safety protocols may be able to differentiate itself in the market, gaining customer confidence and establishing itself as a leader in delivering safe, high-quality food products.

- **Scale and Efficiency:** The competitive merit of an activity can also be assessed based on its capacity to achieve economies of scale and efficiency within the production process. For instance, a company that invests in automation technology for food processing could gain a competitive edge by reducing labor costs, accelerating production rates, and optimizing operational efficiency compared to competitors who may not have adopted such technology.
- **Innovation:** Another crucial factor is evaluating how each activity contributes to driving innovation and the development of new food products and technologies. A company that prioritizes research and development (R&D) may enjoy a competitive advantage by being able to introduce new, innovative food products that set it apart from competitors.
- **Brand Reputation:** It is important to assess how each activity contributes to the establishment and maintenance of the company's brand reputation. A strong brand reputation fosters customer loyalty, reduces the need for extensive marketing, and positions the company more favorably in the market, granting it a competitive edge.
- **Customer Relationships:** Evaluating how each activity contributes to the development and maintenance of customer relationships is also essential. Companies with strong, loyal customer relationships are likely to have a competitive advantage, as this can lead to reduced customer churn, more referrals, and enhanced customer loyalty.
- **Supply Chain Relationships:** Assessing the role of each activity in fostering positive relationships with suppliers and partners throughout the supply chain is key. Companies with strong supplier relationships may gain a competitive advantage by reducing costs, improving product quality, and boosting overall supply chain efficiency.
- **Improving Supply Chain Visibility:** Increasing visibility across the entire supply chain is critical for pinpointing waste and implementing targeted solutions. Companies are leveraging advanced technologies to track and monitor food products as they move through the supply chain. This improved visibility leads to better inventory management, minimizes losses caused by spoilage, and facilitates quicker, data-driven decision-making.
- **Cold Chain Management:** Proper temperature control throughout the cold chain is vital for preventing food spoilage and extending shelf life. To ensure optimal conditions and reduce food waste, companies are investing in state-of-the-art refrigeration technologies, efficient cold storage facilities, and reliable transportation systems that maintain the required temperatures throughout the process.
- **Product and Packaging Innovations:** Companies are increasingly dedicating resources to R&D efforts focused on developing innovative products and

packaging solutions that help minimize waste. Techniques such as modified atmosphere packaging, intelligent packaging, and compostable materials are being used to extend shelf life, reduce spoilage, and support overall sustainability goals.

- **Collaborative Partnerships:** Many companies are building partnerships across the food value chain to collectively tackle food waste. These collaborations involve close coordination between farmers, suppliers, manufacturers, distributors, retailers, and even non-profits. By sharing information, resources, and best practices, these partnerships can streamline processes, reduce inefficiencies, and help minimize food waste across the entire supply chain.
- **Consumer Education and Awareness:** Both companies and governments recognize that consumer behavior plays a vital role in reducing food waste. Initiatives such as the "Love Food Hate Waste" campaign by the UK's Waste and Resources Action Programme (WRAP) and the "Save Food, Share Food" campaign by Feeding India emphasize the importance of educating consumers on reducing waste at home. These campaigns focus on providing practical tips about proper food storage, consumption habits, and overall awareness of food waste issues.

4

Case Study and Analysis of Value Chain of Horticultural Commodities

Agriculture in India, particularly in the northern plains, is undergoing a critical phase. The existing farming practices are becoming unsustainable, prompting the exploration of alternative options for their economic feasibility. Among these alternatives, vegetable crops present a promising choice due to their high value and suitability for multiple cropping systems, offering a viable way to diversify agriculture in Punjab. However, due to their perishable nature and inadequate post-harvest handling, 25-30% of the produce becomes unfit for consumption. This results in higher production costs and lower returns. This chapter explores the potential of the value chain for vegetables and fruits, identifies limitations, and suggests solutions to address the challenges.

The significant losses of fruits and vegetables produced across the country are primarily due to the absence of proper infrastructure for storage and transportation under controlled conditions. Recently, Supply Chain Management (SCM) has gained immense attention due to the effects of globalization. A supply chain involves a series of organizations connected by upstream or downstream flows of products, services, finances, and information from the source to the customer. SCM, therefore, promotes a supply chain approach, focusing on the proactive management and coordination of the two-way movement of goods, services, information, and funds from raw materials to the final consumer. The evolving lifestyle and open economy have pressured manufacturers and suppliers to produce and supply high-quality products.

Various factors are driving the emphasis on supply chain management. First, the availability and cost of information resources between entities within the supply chain enable easy linkages, helping to reduce delays in the network. Second, the heightened competition in both domestic and international markets demands that organizations remain fast, agile, and adaptable. Third, customer expectations and requirements have become mey demanding. To meet these consumer needs, SCM systems must prioritize two main objectives: timeliness and quality.

SCOPE OF SUPPLY CHAIN MANAGEMENT (SCM)

Supply chain management compenises a wide range of activities beyond just purchasing. Each of these activities, though they may seem distinct, is an integral part of a network that dictates how goods and information flow across the supply chain with efficiency and effectiveness. The following are key activities within the supply chain:

- **Purchasing:** Purchasing is a central activity for most organizations, as it focuses on acquiring goods and services that are essential for the functioning of the supply chain.
- **Quality Control:** Most organizations recognize the critical importance of supplier quality and the necessity of preventing quality issues, rather than merely detecting them. Leading organizations collaborate with suppliers to establish robust quality control systems and procedures.
- **Demand and Supply Planning:** Demand planning involves forecasting anticipated demand, managing inventory adjustments, fulfilling unfilled orders, and accounting for spare part and after-market needs. Supply planning, on the other hand, takes these demand forecasts and establishes a network of supply, production, and logistics to meet the required demand.
- **Material or Inventory Control:** The material management team is often tasked with determining the necessary inventory levels of finished goods to meet customer demand. This role highlights the importance of physical distribution, focusing on the outbound or downstream side of the supply chain.
- **Order Processing:** Order processing ensures that customers receive the materials they need, when and where they need them. It acts as a crucial link between the producer and the external customer.
- **Production Planning, Scheduling, and Control:** This area involves creating a time-sensitive production schedule, developing short-term production timelines, and overseeing the management of work-in-process production.
- **Warehousing/Distribution:** Warehousing and distribution are especially critical for companies that produce based on forecasts to meet future sales expectations.
- **Customer Service:** Customer service covers a wide range of activities aimed at ensuring customer satisfaction with both products and services.

Future Potential of Vegetable and Fruit Farming in Punjab

Potato is the dominant vegetable crop in Punjab, which covers an area of 87.2 thousand hectares and accounting for 55% of the vegetable production in the state. There is potential to expand the area dedicated to processing potatoes, as much of the raw material used for processing in Punjab is currently imported from other states. In addition to potatoes, other key vegetables cultivated in Punjab include peas, cucurbits, chilies, capsicum, onions, cauliflower, tomatoes, brinjal, and carrots. The production of turmeric, ginger, and garlic is also gradually increasing. Onion is another key crop, both for domestic consumption and export, and could be promoted as an alternative crop in the state. Pea is the third most significant vegetable crop, consumed both fresh and processed into frozen products, with availability in the state from November to March.

Among fruits, Punjab leads in citrus production, with Kinnow being the dominant variety, occupying 47.1 thousand hectares and contributing to 67% of the state's total fruit production. Other important fruit crops include guava, mango, pear, and peach, which also cover substantial areas in the state. Protected cultivation is an emerging sector with the potential to boost Punjab's economy, creating employment and entrepreneurship opportunities for rural youth. Farmers are participating in vocational training courses organized by the Precision Farming Development Centre at PAU Ludhiana to enhance their skills. Protected cultivation significantly increases land productivity, thereby improving farmers' income levels. Precision farming practices are gaining momentum, with poly net house farming of vegetables becoming popular among both small/marginal farmers and those with larger holdings. Around 1,200 poly net houses, covering 51 hectares, have been established in Punjab. High-value crops like colored and green capsicum, seedless cucumber, tomatoes (table and cherry varieties), gerbera, and lilium are cultivated year-round. Both progressive and small/marginal farmers are adopting these technologies. The Punjab Horticulture Department is reaching out to more farmers to encourage this shift.

Status of Vegetable & Fruit Processing

Although Punjab is a leading agricultural state in India, it lags behind in food processing and value addition. A limited number of low-capacity, multi-commodity processing units are managed by self-help groups. Some self-help groups have also started supplying fresh vegetables directly to retailers and consumers. Establishing a processing industry for bulk consumption of produce would help stabilize vegetable marketing and sustain fruit and vegetable production. Crop cultivars and agronomic practices need to ensure a continuous supply of high-quality raw materials for extended periods.

Establishment of Food Parks

To support small and medium-scale units, food parks are being established to reduce the cost of major facilities such as research and development labs, cold storage, warehousing, and packhouses. Currently, one food park project is underway in Sirhind, district Fatehgarh Sahib, implemented by PAGREXCO. Another food park, approved by the Ministry of Food Processing Industries (MoFPI), is being developed by International Mega Food Park Ltd in village Dhabwala Kalan, district Fazilka. A third food park has also been approved by MoFPI at Ladhowal, Ludhiana.

To support value addition and preservation, MoFPI has approved two cold chain projects in the state. One project at Rajaur, district Kapurthala, is complete and has begun commercial operations, while another at Channo, district Sangrur, is nearing completion.

Developing a Robust Cold Chain Infrastructure

A cold chain is a logistical system designed to maintain optimal storage conditions for perishable goods from the point of origin to the point of consumption across the entire supply chain. The cold chain process begins at the farm level (e.g., harvest methods and pre-cooling) and continues up to the consumer level or at least to the retail level. A well-managed cold chain reduces spoilage, preserves the quality of harvested produce, and ensures a cost-efficient delivery to consumers, with proper attention to customer service.

Advantages of Cold Chain

Advantages for Farmers or Producers

- Increased income for farmers
- Contribution to price stability
- Prevention of glut-like situations

Advantages for Consumers

- Stability in pricing
- Availability of products year-round
- Assurance of consistent product quality

A critical aspect of the cold chain is that if any link is missing or not functioning properly, the entire system collapses. The Cold Chain Logistics Infrastructure typically compeises of:

- Pre-cooling facilities
- Cold storage units
- Refrigerated carriers
- Packaging systems
- Warehousing solutions
- Information management systems (including traceability and tracking)

The cold chain process begins at the farm level and continues until the commodities reach the consumer in good condition.

Present Post-Harvest and Cold Chain Infrastructure for Vegetables and Fruits

- **Cold Storages:** There are currently 550 cold storage units in the state, primarily used for storing potatoes. The Punjab Mandi Board has developed advanced post-harvest infrastructure across 12 fruit and vegetable markets in the state, including 16 multi-chamber cold storage units, each with a 50 MT capacity, and 52 ripening chambers, each with a 10 MT capacity.
- **Mechanical Washing & Grading Lines:** Punjab Agro Industries Corporation (PAIC) has set up five mechanical washing, grading, and waxing plants at various locations including Chhauni Kalan and Kang Mai in Hoshiarpur

district, Abohar, Jattan Tahliwala in Ferozepur district, and Badal in Muktsar district. Additionally, over 100 progressive growers have installed similar facilities on their own farms.

- **Processing Plants:** The Punjab government, through Punjab Agro Juices Limited (PAJL), has established two pilot-scale processing units at Hoshiarpur and Abohar. These plants can process 20 MT per hour of citrus fruits and 10 MT per hour of other fruits and vegetables, and they have cold storage and deep freezer capacities of 1,000 MT at each location. The plants are equipped with machinery procured from CFTSPA Italy, a leading global food engineering company. The plants are designed to process Kinnow and de-bitter the juice, and they can also handle various fruits and vegetables, including tomatoes, carrots, pomegranates, papayas, guavas, pears, pumpkins, melons, chilies, jamun, mangoes, strawberries, litchis, amla, and aloe vera. They posess capacity to process pulp processing pulp and storing concentrates and single-strength juices at the same facility. PAJL also performs job work for large domestic and multinational companies.
- **Centre of Excellence:** To help farmers enhance their skills and stay updated with the latest agricultural practices, the Punjab government has established three Centers of Excellence. One focuses on vegetables in Kartarpur (Jalandhar), another on fruits in Khanaura (Hoshiarpur), and the third on potatoes in Dhogri (Jalandhar). The center at Kartarpur was set up in collaboration with the Indo-Israel agreement and aims to introduce Punjab's farmers to cutting-edge technology and scientific advancements in vegetable cultivation. The Dhogri center, developed in partnership with the Indo-Dutch joint agricultural workgroup, focuses on improving seed potato quality, suitable farm mechanization, and disease management technologies.
- **High-Tech Pack Houses:** Five high-tech pack houses equipped with cold storage, grading, and sorting lines have been established by PAGREXCO at various locations, including Mushkabad (Ludhiana), Saholi and Lalgarh (Patiala), Kangmai (Hoshiarpur), and Babri (Gurdaspur).
- **Self-Help Groups:** Cooperative societies such as FAPRO, Unati, and KAFRO are currently serving as role models for the development of rural entrepreneurs. These groups contribute to income generation and employment creation in rural areas.

Growth of Private Sectors in Punjab

- **Field Fresh Foods Private Ltd:** This is a joint venture between Bharti Enterprises and Del Monte Pacific Limited. For fresh vegetables, Field Fresh Foods operates an Agriculture Centre of Excellence (ACE) at Ladhowal, Ludhiana, Punjab. This facility is one of the largest agricultural R&D centers of its kind, covering 300 acres, with state-of-the-art protected and open field cultivation methods for value-added vegetables. The ACE showcases advanced technologies and best agricultural practices to partner farmers, helping them improve productivity and farm income sustainably. Field Fresh Foods collaborates with over 4,000 partner farmers across 5,000 acres in Punjab and Maharashtra. These farmers are Global GAP compliant and grow vegetables for export to Europe. The company sells processed food and beverages in India under the Del Monte brand and fresh vegetables in export markets under the Field Fresh brand.

- **Namdhari Farm Fresh:** Namdhari's Fresh processes over a thousand tons of fresh vegetables and fruits for both domestic and international markets, with exports primarily to Europe, Australia, and the Middle East. After their success in the seed industry, Namdhari's has expanded into the production, distribution, and export of fresh vegetables and fruits under the "Namdhari's Fresh" brand. The company maintains a continuous cold chain network to ensure the freshness of the produce until it reaches consumers. After harvesting, vegetables are transported in refrigerated trucks, which help remove field heat. From the refrigerated trucks, they are moved to a pre-cooling room. The company is certified by the Dutch accredited certifying agency SKAL, with the EKO symbol, recognized in Europe and North America. Namdhari's is the first vegetable-growing and exporting firm in India to receive the GLOBAL-GAP certification. The pack house is in the process of implementing BRC and HACCP, and all necessary practices and norms are already in place.
- **PAGRO Frozen Foods Private Ltd:** PAGRO Frozen Foods, established in 2007, is the largest private integrated vegetable processing plant in Punjab. The primary product is frozen green peas, though it also processes other vegetables such as potatoes, onions, carrots, and sweet corn. With a daily processing capacity of 550 tons (50,000 tons of vegetables per year), PAGRO is the largest plant of its kind in South East Asia. The plant operates for 280 days annually and sources approximately 90% of its raw material from outside the state.

TOMATO PROCESSING AND ITS MARKET POTENTIAL

The World Processing Tomato Council (WPTC) reported that in 2015, approximately 41.37 million tonnes (MT) of tomatoes, accounting for 26% of global production, were processed into value-added products worldwide. In contrast, India processed less than 1% of its total tomato production into these products, with only an estimated 130,000 tonnes processed in 2015—just 0.3% of the global market for tomato processing. Tomato paste is a key product in this industry, as it serves as the base for many other tomato-based products. These include tomato juice, paste, diced or peeled tomatoes, strained pulp, ketchup, pasta sauces, pickles, pizza sauces, salsa, gravies, ready-to-eat (RTE) curries, and tomato-based powders. Processed tomato products are widely used across various sectors, including households, food processing industries, snack food companies, hotels, restaurants, and fast food chains. The processing typically involves washing and grading the tomatoes, followed by steaming them in steam-jacketed kettles and pulping in a continuous pulper, which separates the skin and seeds from the pulp. The filtered pulp is then used as the raw material for creating other products. The yield of pulp generally ranges from 40% to 50%, depending on the quality of the tomatoes. According to a recent survey by ASSOCHAM (Associated Chambers of Commerce of India), demand for tomato puree and ketchup saw a 40% increase in

June 2016, driven by rising fresh tomato prices and a shift toward easier-to-prepare meals. India's annual ketchup consumption is estimated at 13,000 tonnes, with a market value of Rs 1.8 billion (USD 28 million). Nestle's Maggi holds the largest share of the ketchup market at 37%, followed by Unilever's Kissan at 29% and Heinz at 10%. Producing tomato paste appears commercially viable when processing facilities are situated near major tomato-growing areas in India. However, despite the expanding market for processed tomato products, the processing industry often faces challenges related to the limited supply of tomatoes suitable for processing. Essential quality parameters for processing, such as color, soluble solids, sugar content, and firmness, are not met by the existing Indian tomato varieties, which are more suited for fresh consumption. The varieties most favored in the fresh market are acidic and ideal for curries and common dishes, but processors must neutralize this acidity by adding more sugar, thus increasing production costs. Despite the potential for high tomato production, low yields raise both costs and risks for farmers, which ultimately results in reduced incomes for those involved in tomato farming.

Leading Indian Tomato Paste Producers and Manufacturers

- **Hindustan Unilever:** HUL's Kissan brand, with a 25% market share, is the second-most popular ketchup brand in India. In 2011, HUL pioneered farm-gate sourcing of tomatoes from smallholder farmers in Nasik district. It partnered with local tomato paste processor Varun Agro and agri-input suppliers to create a local supply chain for producing tomato paste for ketchup production. In 2011, Unilever sourced 40,000 tonnes of tomatoes from India, making up 60% of its tomato paste requirement for ketchup.
- **Field Fresh Foods (Del Monte):** The third-largest processed tomato product manufacturer in India, Field Fresh produces ketchup, pasta, and pizza sauce under the Del Monte brand. Its factory is based in Krishnagiri, and the company has set up a 120-hectare R&D farm in Ladhowal, near Ludhiana, where it tests specialized cultivars for processing and mechanization in tomato cultivation and harvesting.
- **Nestle India:** India's top ketchup maker, Nestle controls a 37% market share through its Maggi brand. In addition to ketchup, Nestlé produces tomato soup mixes to compete with brands like Knorr. The company collaborates with local suppliers under its Supplier Development Program but also imports raw materials globally for its ketchup production in India.
- **Global Green:** Based in Bangalore, Global Green began large-scale contract farming of hybrid tomatoes in 2012 to meet the rising demand for tomato paste and sauces. The company promotes the use of hybrid varieties from United Genetics USA, which are rich in lycopene, and sources tomatoes from Andhra Pradesh, Kolar, and neighboring areas of Karnataka. The processed tomatoes are turned into value-added products under the Tify brand, including ketchup, pasta sauce, and pizza sauce. The company processes over 20,000 tonnes of tomatoes annually.

- **Indira Foods:** Founded in 2008, Indira Foods produces tomato ketchup for both retail and institutional markets, primarily focusing on southern India. With a daily processing capacity of 30 tonnes, the company commands 70% of the hotel, restaurant, and airline market share in Karnataka and supplies to Global Green and Namdhari Group. Tomatoes are sourced from Kolar district.
- **Cremica Group:** Cremica produces a variety of tomato products, including ketchup, puree, pasta sauce, dips, and Indian gravies at its production plants in Ludhiana (Punjab) and Noida (near Delhi). The company supplies food chains such as McDonald's, Barista, Café Coffee Day, Pizza Hut, Domino's, and Papa John's, as well as institutional partners like Big Bazaar, Spencer's, Taj Group, ITC, Jet Airways, and Air India.
- **Dabur India:** A leading producer of tomato puree, tomato juice, soups, and chutneys under its 'Homemade' brand, Dabur commenced its tomato processing operations in 2011 at its Siliguri (West Bengal) plant. It also produces juices at its factory in Nepal.
- **Capricorn Food Products:** One of India's largest independent tomato paste manufacturers, Capricorn supplies paste and puree to major processors like Hindustan Unilever, Nestle, and Field Fresh. Its processing plants in Nashik (Maharashtra), Koyna (Maharashtra), Krishnagiri (Tamil Nadu), and Chitoor (Andhra Pradesh) are operational. In 2013, Capricorn opened a plant in Nashik with a daily capacity of 100 MT of tomato paste. During the off-season, it processes mango pulp to maintain year-round production.
- **Nijjer Agro Foods:** Based in Amritsar, Nijjer Agro Foods produces tomato paste for clients such as Nestle, Del Monte, and other processed food manufacturers in northern India.
- **GD Foods:** Operating in Tarn Taran (Punjab), GD Foods manufactures tomato paste and processed products like ketchup under its 'Tops' brand. With a production capacity of 42,000 tonnes annually, the company also processes chili and apple products. GD Foods works with farmers on 400 hectares through contract farming.
- **Mother Dairy:** Located in Karnataka, Mother Dairy produces tomato paste and its Safal brand of ketchup. The company partnered with Bayer Crop Science to improve tomato production among farmers in Chickballapur and Tumkur districts, leading to an increase in tomato yields from 35 tonnes per hectare to 45 tonnes per hectare. In 2016, it announced plans to set up a new puree production unit in Ranchi, Eastern India. The company also sells frozen tomatoes in the Delhi-National Capital Region, processed at its West Delhi facility.
- **Godrej Beverages and Foods:** Godrej markets its Smart Cook tomato puree range for the home cooking segment.
- **GRG Foods:** Based in Bangalore, GRG Foods produces the Spego brand of tomato ketchup and Revathi brand of tomato-based powders and mixes for the Southern India market.

- **National Agriculture Co-operative Marketing Federation (NAFED):** NAFED operates a tomato paste and ketchup production plant in Vellore (Tamil Nadu).
- **ITC Group:** ITC manufactures a range of ready-to-eat (RTE) products under the Kitchens of India brand for the Indian and export markets. These include curry pastes, sauces, and chutneys, with tomato as a key ingredient.
- **Griffith Laboratories:** A U.S.-based condiment producer, Griffith manufactures Indian paste and powder mixes at its Bengaluru facility for institutional clients, including hotels and restaurant chains in India and internationally.
- **Chordia Food Products:** Operating in Shirwal (Maharashtra), Chennai, and Dharwad (Karnataka), Chordia produces tomato ketchup, paste, and mixes for both the domestic retail market and institutional clients. With a processing capacity of 2.5 tonnes per day, the company previously supplied tomato paste to Nestle but now primarily consumes its processed paste.

Challenges and Limitations

The following are key constraints that discourage farmers from growing vegetables and fruits for processing:

- The quality of fruits and vegetables is inconsistent.
- There is a lack of processing units in key crop production areas.
- Seeds from private companies are expensive.
- The domestic demand for processed products remains low due to the year-round availability of most fresh fruits and vegetables.
- Manpower involved in logistics and marketing often lacks knowledge about produce requirements, leading to quality and value losses.
- There is an absence of contract farming systems.
- Processing units often delay payments to growers.
- Processing units frequently treat farmers poorly.
- There is insufficient mechanization in crop cultivation and handling processes.
- The varieties of crops available are often unsuitable for processing purposes.
- Disease and pest issues impact crops.
- Power supply is unreliable and expensive.
- Pricing offered by processors is inadequate.
- Rural infrastructure is poor, affecting transportation.
- The long-term viability of the processing industry remains uncertain for various reasons.

Essential Remedies and Strategies

- Simply increasing the area and productivity of vegetables will not address Punjab's diversification and value chain challenges unless all farmers are integrated into a modern value chain structure.
- Periodic gluts of specific fruits and vegetables can be managed by establishing proper post-harvest infrastructure, including cold chain facilities, both at farm and market levels.

- If middlemen are involved in the supply chain, they should act as facilitators, adding value by consolidating produce, engaging in wholesale activities, and supporting processing, storage, packaging, and transportation.
- Strengthening the linkages between farmers and food processors is essential.
- Innovative approaches, such as contract farming, should be encouraged. This can help by providing farmers with quality inputs, technical guidance, and management skills.
- To access high-value markets, there is a need to develop additional skills and expertise among farmers.
- To promote agricultural diversification, Punjab must prioritize research and development, as well as extension and training services in post-harvest management, distribution, and marketing of horticultural crops.

5

Sourcing and Material Management

Sourcing and material management in agriculture indicates the process of identifying and managing suppliers and materials to maintain a steady flow of high-quality products. It also ensures that sourcing is both sustainable and cost-effective. Sourcing management plays a key role in an organization's supply chain, ensuring that a company can obtain raw materials, tools, and other resources needed to deliver its products and services at the best possible cost. However, sourcing management is more than just finding reliable suppliers. It involves multiple layers, such as evaluating the quality and dependability of suppliers, maintaining strong supplier relationships, negotiating prices, optimizing costs, and addressing potential supply chain risks—both known and unknown. Let's dive into how sourcing management functions and its significant impact on a company's profitability, the quality of its products and services, and its overall resilience.

DEFINING SOURCING MANAGEMENT

Sourcing management is the practice of identifying suppliers who can provide the necessary resources for a business at the most cost-effective rate without compromising on quality or reliability.

Understanding Differences Between Sourcing and Procurement

Although sourcing and procurement are closely linked, they each serve distinct business functions and offer different advantages:

- **Sourcing** involves identifying ideal suppliers, negotiating contracts and prices, and managing relationships with these suppliers. Sourcing management uses procurement data to evaluate supplier performance and make strategic decisions that either reduce costs or maintain a consistent supply chain.
- **Procurement** focuses on acquiring the materials, tools, and services required to produce and deliver a company's products or services. Operating within

the supply chains established by sourcing, procurement handles ordering, tracking deliveries, and managing budgets based on the prices negotiated through sourcing.

While both processes are intertwined and rely on each other, everything begins with strategic sourcing management.

WORKING OF STRATEGIC SOURCING MANAGEMENT

The process of strategic sourcing management involves several key steps:

- Identifying a company's needs and expectations for raw materials, products, and services
- Conducting market research to discover potential suppliers
- Comparing current and prospective suppliers
- Identifying the best suppliers based on quality and cost
- Recognizing backup suppliers to ensure reliability
- Negotiating terms, including contracts, minimum order quantities, payment terms, and pricing
- Assessing and managing supplier risks
- Continuously monitoring supplier performance to maintain quality

Overall, sourcing management is focused on selecting the most reliable suppliers who offer the best products or materials at the lowest price. Companies are careful not to compromise on the quality of raw materials for a lower cost, as doing so could degrade the quality of the company's products or services, leading to customer loss, damaged reputation, and negatively affecting the company's financial performance.

EMBRACING SUPPLIER DIVERSITY IN AGRICULTURE IN INDIA

In order to promote inclusive growth, improve economic sustainability, and guarantee fair opportunities for underprivileged communities in India, supplier diversity in agriculture is a crucial first step. The idea is to incorporate a variety of suppliers into the agricultural supply chain, such as small-scale farmers, women-led businesses, tribal cooperatives, and rural entrepreneurs. Accepting supplier diversity can have a big impact on rural development, innovation, and food security in India because of the country's varied agricultural landscape and socioeconomic inequalities.

Importance of Supplier Diversity in Indian Agriculture

For small and marginalized farmers, who are the backbone of Indian agriculture, to have access to bigger markets and equitable pricing structures, supplier diversity in the sector is essential. Economic distress frequently results from smallholder farmers' limited opportunities due to the dominance of large agribusiness corporations and middlemen. Businesses and policymakers can establish an inclusive ecosystem where farmers from a variety of backgrounds actively participate in the supply chain by encouraging supplier diversity.

Diverse suppliers also promote the use of sustainable agricultural methods. Indigenous communities and smallholder farmers frequently use eco-friendly, traditional farming practices that complement contemporary sustainability objectives. Promoting organic and regenerative farming methods, lowering environmental impact, and protecting biodiversity are all aided by encouraging their involvement.

Challenges in Implementing Supplier Diversity

Implementing supplier diversity in Indian agriculture is fraught with difficulties, despite its benefits. Lack of resources and credit availability for small farmers, especially those from underserved communities, is a significant barrier. Their participation in mainstream agricultural markets is limited by their lack of financial literacy and difficulties with land ownership.

The prevalence of middlemen in the supply chain presents another difficulty. Farmers frequently depend on intermediaries, who take advantage of their inexperience in the market to offer lower prices for their produce. It is challenging for small suppliers to compete when they do not have direct access to large buyers. Furthermore, the integration of diverse suppliers into mainstream agriculture is hampered by infrastructural shortcomings like insufficient cold storage, transportation, and supply chain inefficiencies. Small farmers are also unable to take advantage of opportunities due to a lack of knowledge about government programs and a lack of digital literacy.

Strategies to Promote Supplier Diversity

To enhance supplier diversity in Indian agriculture, various strategies can be adopted:

1. Government Support & Policy Reforms: bolstering laws that provide financial aid, training initiatives, and subsidies to women farmers, cooperatives, and smallholder farmers.
2. Technology & Digital Inclusion: Using digital marketplaces and e-commerce platforms to link a variety of suppliers and buyers directly, cutting out intermediaries.
3. Corporate Responsibility: Promoting the practice of large retailers and agribusinesses obtaining a portion of their agricultural products from diverse and small suppliers.
4. Infrastructure Development: To guarantee small farmers' seamless integration into the supply chain, investments should be made in cold storage, rural roads, and logistics.
5. Capacity Building & Training: To increase the competitiveness of diverse suppliers, training in sustainable farming, financial literacy, and market access is provided.

India can build a more robust and inclusive agricultural sector that benefits farmers, companies, and consumers by embracing supplier diversity.

STRATEGIES FOR SUPPLIER SELECTION

When it comes to sourcing agricultural products, choosing the right suppliers is essential for ensuring the quality and reliability of the products we provide. The supplier selection process plays a pivotal role in determining the success of our agricultural operations.

To identify and collaborate with the right suppliers, it is important to adopt effective supplier selection strategies. These strategies guide us in making well-informed decisions and establishing long-term partnerships with suppliers who are aligned with our goals and standards.

Comprehending the Process of Supplier Selection

The supplier selection process involves several critical steps that help us assess and select the most suitable suppliers for our agricultural needs. The first step is defining our requirements and specifications, taking into account factors such as product quality, quantity, pricing, delivery capabilities, and sustainability practices.

Once we have a clear understanding of our requirements, we move into the supplier evaluation phase. This involves conducting thorough research, reviewing supplier portfolios, and collecting relevant data to assess their capabilities and reputation within the industry.

After this initial evaluation, we can shortlist a few potential suppliers and begin negotiations and collaboration. This stage involves discussing key elements such as terms and conditions, pricing, delivery schedules, and any other specific requirements that may arise.

Throughout the entire process, it is crucial to engage key stakeholders and seek their input to ensure that the supplier selection aligns with the organization's overall objectives and values.

Effective Supplier Selection Strategies

To ensure effective supplier selection, below are some strategies to keep in mind:

- **Define Clear Selection Criteria:** It's important to establish clear selection criteria that match your specific agricultural needs. This includes factors such as product quality, reliability, cost-effectiveness, sustainable practices, and timely delivery.
- **Conduct Thorough Research:** Gather comprehensive information about potential suppliers by utilizing online research, industry publications, and recommendations from trusted sources. This will provide valuable insights into their reputation, customer feedback, and track record in the agricultural sector.
- **Carry Out Detailed Evaluations:** Evaluate potential suppliers through various methods, such as interviews, site visits, product samples, and reference checks. This will help you assess their capabilities, quality control measures, and adherence to industry standards.

- **Focus on Long-Term Relationships:** Prioritize suppliers who are open to building long-term partnerships. Such collaborations can lead to enhanced reliability, improved cooperation, and mutual growth over time.
- **Evaluate Financial Stability:** Assess the financial health of potential suppliers to ensure they can consistently meet your demands, particularly during periods of market volatility or economic uncertainty.

By incorporating these supplier selection strategies, we can strengthen the agricultural supply chain and ensure a continuous and reliable flow of top-quality agricultural products.

"The quality of our agricultural products is directly linked to the suppliers we choose. Implementing effective supplier selection strategies is vital for our success and the overall development of the agricultural industry."

Proven Methods for Finding Agricultural Product Suppliers

Finding dependable suppliers is essential to ensuring a consistent and diverse supply of high-quality agricultural products. To help you in this journey, below are some proven methods and techniques to identify the right suppliers for your agricultural needs:

- **Networking:** Networking is a crucial tool in connecting with potential agricultural product suppliers. Attend agricultural trade shows, industry conferences, and local farmers' markets to meet various suppliers. Engaging in these events allows you to foster relationships, make meaningful connections, and open the door for future partnerships.
- **Online Directories:** Make use of online directories dedicated to agricultural product suppliers. These platforms offer comprehensive details about suppliers, including their contact information and the range of products they provide.
- **Local Cooperative Associations:** Consider joining local cooperative associations that unite farmers, producers, and suppliers. These organizations offer a platform for collaboration, providing ample opportunities to connect with suppliers in your region. By becoming a member, you can expand your network and access a range of agricultural products.
- **Farmer-to-Farmer Networks:** Engage with farmer-to-farmer networks where experienced farmers share insights about suppliers they trust. These networks are invaluable for discovering suppliers with a proven reputation for delivering quality agricultural products on time.
- **Online Marketplaces:** Take advantage of online marketplaces tailored to agricultural products. Platforms like insert relevant marketplaces allow you to browse through a wide variety of suppliers, compare pricing, and read reviews from other customers, helping you make well-informed decisions when selecting suppliers.

By leveraging these methods, you can find reliable agricultural product suppliers that meet your specific needs and contribute to the success of your agricultural business.

"Finding reliable agricultural product suppliers is crucial for ensuring a steady and diverse supply of high-quality goods."

Remember, finding the right agricultural product suppliers is an ongoing process. Continuously evaluate and assess your suppliers' performance, considering factors like quality, reliability, and consistency.

Best Practices for Agricultural Product Sourcing

Efficient sourcing of agricultural products is crucial for ensuring a steady, high-quality, and cost-effective supply of goods to meet consumer demand. By adopting best practices in agricultural product sourcing, businesses can optimize their operations, reduce risks, and build long-term relationships with reliable suppliers. Below are the key best practices that can enhance agricultural product sourcing:

Clearly Define Sourcing Requirements

- **What to Do:** Before you begin sourcing agricultural products, clearly define the specifications, quality standards, quantity, and delivery requirements. Understanding your needs helps in selecting the right suppliers who can meet your expectations.
- **Why It Matters:** Clear requirements ensure that suppliers understand your expectations and can meet them without ambiguity. This reduces the risk of receiving subpar or inadequate products.

Establish Robust Supplier Connections

- **What to Do:** Develop long-term Connections with suppliers by fostering trust and mutual collaboration. Communicate regularly with suppliers, offer feedback, and maintain transparency regarding any changes in demand or requirements.
- **Why It Matters:** Strong relationships help build reliability, ensure timely deliveries, and facilitate smoother communication. Over time, these relationships lead to better terms, such as discounts, priority deliveries, or flexibility in case of emergencies.

Diversify Your Supplier Base

- **What to Do:** Avoid depending on a single supplier for your agricultural products. Work with multiple suppliers to mitigate the risk of disruptions in supply caused by weather conditions, supply chain issues, or other unforeseen events.
- **Why It Matters:** Diversifying your supplier base helps ensure a consistent supply of products. It protects your business from disruptions and provides more options if one supplier cannot meet your needs.

Conduct Thorough Supplier Evaluation

- **What to Do:** Evaluate potential suppliers based on various factors such as product quality, production capacity, reliability, delivery timelines, pricing, and sustainability practices. It's important to gather information from multiple sources, including reviews, referrals, and audits.

- **Why It Matters:** A comprehensive evaluation of suppliers ensures that you select only the most reliable and capable partners. This helps maintain product quality, manage risk, and optimize the supply chain.

Implement Quality Control Procedures

- **What to Do:** Set up strict quality control protocols for the products you source. This could involve inspections, certifications, lab testing, or audits. Ensure that suppliers adhere to industry standards and regulations.
- **Why It Matters:** Maintaining high-quality standards ensures that the products you source are safe for consumption, meet regulations, and maintain consumer trust. It helps prevent costly recalls or reputational damage due to poor quality products.

Use Technology and Data for Sourcing Decisions

- **What to Do:** Leverage technology such as digital platforms, supply chain management software, and data analytics to streamline sourcing operations. Use these tools for demand forecasting, tracking supplier performance, and analyzing procurement data.
- **Why It Matters:** Technology allows for more efficient sourcing, accurate decision-making, and better management of suppliers and inventory. Data-driven insights help optimize procurement costs, reduce waste, and improve overall supply chain efficiency.

Negotiate Terms and Prices Effectively

- **What to Do:** Always negotiate with suppliers to get the best possible terms. While price is a key factor, don't overlook other elements such as payment terms, delivery schedules, and return policies. Negotiate in a way that benefits both parties.
- **Why It Matters:** Effective negotiations ensure cost savings, more favorable payment terms, and better service from suppliers. It also sets the tone for a professional, long-term partnership that benefits both parties.

Monitor and Manage Supplier Performance

- **What to Do:** Regularly monitor supplier performance in terms of quality, delivery schedules, and customer service. Set performance metrics and track the supplier's ability to meet these standards over time.
- **Why It Matters:** Monitoring performance helps identify and address issues before they escalate. It ensures that suppliers continue to meet your expectations and improves the reliability of your sourcing process.

Promote Sustainability and Ethical Sourcing

- **What to Do:** Incorporate sustainability and ethical practices into your sourcing strategy. Work with suppliers who use environmentally friendly practices, support fair trade, and prioritize the well-being of their workers and communities.
- **Why It Matters:** Sustainable sourcing helps reduce environmental impact, supports ethical labor practices, and improves the social responsibility of your business. It can also appeal to socially conscious consumers who value sustainability.

Be Prepared for Supply Chain Disruptions

- **What to Do:** Develop contingency plans in case of supply chain disruptions. Identify backup suppliers, stockpile essential products when necessary, and ensure flexibility in your logistics and transportation.
- **Why It Matters:** Being prepared for disruptions ensures that your business can continue operating smoothly in case of natural disasters, political unrest, or other emergencies. It helps maintain business continuity and consumer satisfaction.

Ensure Compliance with Local and International Regulations

- **What to Do:** Stay updated with local, national, and international regulations regarding agricultural products. Ensure that all sourced products comply with food safety standards, packaging regulations, and import/export laws.
- **Why It Matters:** Compliance with regulations is crucial for ensuring the legality and safety of your products. Non-compliance can result in legal penalties, fines, or product recalls that can damage your brand reputation.

Prioritize Transparent Communication

- **What to Do:** Maintain clear and open communication with all stakeholders in the sourcing process, including suppliers, distributors, and internal teams. Address any concerns or issues promptly and keep everyone informed about changes in product specifications, delivery times, or other important details.
- **Why It Matters:** Transparent communication reduces misunderstandings and ensures everyone involved is on the same page. It fosters trust and helps create long-lasting, effective relationships with suppliers.

By following these best practices, businesses can optimize their agricultural product sourcing, reduce risks, and improve the efficiency and reliability of their supply chains. In a competitive agricultural market, adopting a proactive and strategic approach to sourcing will help businesses not only meet consumer demands but also drive growth, sustainability, and resilience in the long term.

Improving Strategies for Agricultural Procurement

To ensure a seamless and efficient agricultural supply chain, implementing well-planned procurement strategies is essential. By enhancing our agricultural procurement processes, we can strengthen supplier management and streamline operations.

Cultivate Strong Supplier Relationships

Building strong relationships with agricultural suppliers is fundamental to successful procurement. Establishing trust and maintaining open lines of communication with suppliers allows us to form long-term partnerships that are mutually beneficial. Regular interactions with suppliers enable us to understand their capabilities, challenges, and needs, helping us align our procurement strategies accordingly. This alignment fosters a more effective and efficient supply chain.

Leverage Technology for Streamlined Procurement

In the digital era, utilizing technology can significantly enhance agricultural procurement. Adopting procurement software or digital platforms can streamline the management of suppliers, automate purchasing processes, and provide valuable insights through data analysis. Embracing these technological solutions will increase procurement transparency, reduce manual errors, and allow for data-driven decision-making to optimize strategies.

Expand the Supplier Base

Diversifying the supplier base is crucial to minimizing risks in agricultural procurement. By working with a variety of suppliers, we can ensure a reliable supply of agricultural products, even when faced with unforeseen disruptions such as natural disasters or market volatility. Supplier diversification also fosters competition, which can lead to better pricing and higher quality options.

PROCUREMENT MANAGEMENT

Procurement management is an important aspect of supply chain operations, ensuring that businesses acquire the necessary goods and services efficiently and cost-effectively. It involves planning, selecting suppliers, negotiating contracts, and overseeing the procurement process to guarantee timely delivery and quality assurance. In the agricultural sector, effective procurement management is essential for maintaining a steady supply of high-quality produce while minimizing costs and reducing risks associated with supply chain disruptions.

A well-structured procurement management system involves strategic sourcing, where businesses evaluate multiple suppliers based on factors such as reliability, cost, product quality, and sustainability practices. Establishing long-term partnerships with trusted suppliers can enhance supply chain stability and foster mutual growth. Additionally, diversification of suppliers reduces dependency on a single source, mitigating potential risks caused by market fluctuations, climate changes, or logistical challenges.

Technology plays a significant role in modern procurement management by streamlining processes and improving efficiency. Digital procurement platforms, data analytics, and automation tools help businesses track supplier performance, optimize inventory management, and forecast demand more accurately. By leveraging these technological solutions, companies can reduce manual errors, enhance transparency, and make data-driven procurement decisions that improve overall efficiency.

Ultimately, effective procurement management contributes to a more resilient and sustainable supply chain. By implementing best practices such as strategic supplier selection, adopting technology-driven solutions, and continuously evaluating procurement processes, businesses can optimize costs, ensure consistent product quality, and strengthen their market position.

Think of procurement management as the skillful process of acquiring the necessary supplies and services to keep your agricultural business operating efficiently. Simply put, it involves sourcing, purchasing, and overseeing essential resources—ranging from seeds and fertilizers to farming machinery and transportation services.

Consider a farmer preparing for the planting season who requires high-quality seeds, fertilizers, and irrigation equipment. Procurement management ensures that the farmer identifies reliable suppliers, negotiates favorable terms, and obtains the necessary materials on time to guarantee a productive harvest. It's all about being strategic—ensuring timely access to essential resources at the best possible cost while maintaining quality and efficiency.

Challenges in Agricultural Procurement

Managing procurement in agriculture is no easy task—it's more like navigating a constantly shifting landscape filled with obstacles. below are some of the most pressing challenges that agribusinesses often encounter:

Price Fluctuations

One of the biggest hurdles in agricultural procurement is dealing with unpredictable price changes. Factors such as weather conditions, global market trends, and geopolitical events can significantly impact the cost of essential supplies like seeds, fertilizers, and farming equipment.

For instance, a severe drought in a key agricultural region may cause a surge in the price of certain crops, making their seeds more expensive. Similarly, a political crisis in a country that exports fertilizers could disrupt supply chains and drive up costs. One moment, a farmer might be purchasing fertilizer at a reasonable price, and the next, prices could double due to circumstances beyond their control. This volatility makes it difficult for farmers—especially those with tight profit margins—to budget effectively and maintain financial stability.

Quality Assurance

Securing high-quality agricultural inputs is another major challenge. Unlike in other industries where defective products can be easily replaced, poor-quality farming inputs can lead to an entire season's failure, resulting in massive financial losses.

For example, if a farmer unknowingly plants low-quality seeds, the yield may be significantly reduced, leading to fewer crops available for sale. In a worst-case scenario, the crop may be so weak that it never reaches harvest. This directly affects the farmer's income and ability to cover operational costs. As a result, finding reliable suppliers that provide high-quality inputs is critical—but far from simple.

Digital Solutions: The Future of Procurement Management in Agriculture

In today's fast-evolving world, agriculture is undergoing a digital transformation, revolutionizing the procurement process. Below are the key digital solutions that are reshaping how agribusinesses manage their purchasing and supply chains.

Automation and Data Analytics

Imagine eliminating the hassle of manually updating supplier lists, tracking inventory, or handling purchase orders. Automation makes this possible by streamlining repetitive procurement tasks, allowing farmers and procurement managers to focus on scaling their businesses rather than being bogged down by administrative work.

Beyond automation, data analytics serves as a powerful tool for making informed procurement decisions. By analyzing past purchasing patterns and market trends, agribusinesses can anticipate price fluctuations and optimize their buying strategy.

Blockchain for Transparency and Traceability

Blockchain technology, often associated with cryptocurrencies, is making a significant impact on agriculture by improving transparency and traceability. Think of it as a digital record-keeping system that securely tracks every stage of a product's journey—from the farm to the final consumer. This technology ensures that every transaction and movement within the supply chain is verifiable, reducing fraud, preventing counterfeit products, and boosting trust among all stakeholders.

Blockchain-powered traceability systems provide farmers and suppliers with a way to verify product authenticity, ensuring that consumers receive high-quality, ethically sourced goods.

Driving Efficiency Through Digital Innovation

The adoption of digital solutions such as automation, data analytics, and blockchain is transforming agricultural procurement by improving efficiency, accuracy, and trust. As agribusinesses continue to embrace these technologies, they will be better positioned to optimize procurement processes, reduce costs, and meet the growing demands of a rapidly evolving global food supply chain.

Imagine you're a buyer searching for organic coffee beans. With blockchain technology, you can track those beans all the way back to the exact farm where they were grown. This system provides detailed information about the entire supply chain, including where the beans were harvested, how they were processed, and whether the farm adheres to organic farming standards.

This level of transparency ensures that you're not just relying on a label but have verifiable proof that the product is genuinely organic. It's a game-changer for buyers who want authenticity and quality, and it also benefits farmers by helping them build trust with customers. With blockchain, both parties win—the buyer gains confidence in their purchase, while the farmer strengthens their reputation in the market.

Best Practices for Procurement Management

Effective procurement management has the power to revolutionize how businesses acquire materials, oversee supplier relationships, and promote sustainability. Below are some best practices, explained in a straightforward and engaging way:

- **Develop Strong Supplier Partnerships:** Think of your suppliers as long-term business partners rather than just vendors. Open communication and mutual trust are essential. By keeping your suppliers informed about your needs, schedules, and quality expectations, you create a smooth workflow that minimizes confusion and ensures consistency in product delivery.
- **Utilize Data for Smarter Decisions:** Data is one of the most powerful tools in procurement. By analyzing past purchasing trends and market fluctuations, businesses can make well-informed decisions. For instance, if historical data shows that a specific product's price usually decreases at a particular time of year, buying in bulk during that period can lead to significant cost savings.
- **Prioritize Sustainability:** Sustainability is no longer just a passing trend; it has become a fundamental expectation from both consumers and businesses. Incorporating sustainable sourcing practices not only helps protect the environment but also enhances a company's reputation. It can attract eco-conscious buyers and create new business opportunities, giving companies a competitive edge in the market.

Advantages of Effective Procurement Management in Agriculture

A well-structured procurement system in agriculture offers numerous benefits, from reducing costs and enhancing product quality to ensuring compliance with evolving regulations. Here's how a strong procurement strategy can positively impact agribusinesses:

Cost Efficiency

Effective procurement management can generate substantial cost savings for agricultural businesses. One of the most impactful ways to achieve this is through bulk purchasing. When buying essential inputs like seeds, fertilizers, and farming equipment in large quantities, suppliers often provide discounts, helping businesses cut down on expenses.

According to a World Bank report, purchasing in bulk can lower costs by 10-20%, depending on order size and market conditions. Additionally, strong supplier relationships facilitate better negotiations, leading to advantageous terms such as extended payment periods or volume-based discounts. These strategies improve cash flow and minimize procurement costs, allowing agribusinesses to allocate resources more efficiently.

Enhanced Product Quality

In agriculture, the quality of inputs directly influences the final yield. A well-organized procurement process ensures that only superior seeds, fertilizers, and other essential materials are sourced, resulting in higher crop productivity and profitability.

By prioritizing premium agricultural inputs, farmers can maximize output without increasing land, labor, or water usage. This not only strengthens profitability but also ensures a steady supply of produce to meet market demands consistently.

Regulatory Compliance

With agricultural regulations becoming more stringent, businesses must proactively ensure compliance with sustainability and traceability standards.

A well-managed procurement system simplifies regulatory compliance by maintaining detailed records of product sourcing. This enables agribusinesses to verify that their inputs align with necessary standards, reducing the risk of penalties, import restrictions, or loss of access to valuable international markets. By integrating compliance into procurement strategies, agricultural businesses can safeguard their operations and maintain their competitive edge.

HANDLING MATERIALS IN AGRICULTURE

Material handling in this context refers to all the processes involved in the movement of agricultural products, whether they are in their raw, in-process, or processed states.

Figure: 5.1 Movement of Grains in a Screw Conveyor

The movement typically occurs over short distances and encompasses activities such as loading, unloading, packing, and distribution. These processes play a crucial role in the efficient management of agricultural products from the farm to consumers, ensuring that goods are handled safely, cost-effectively, and with minimal damage or waste.

Let's use the example of grains stored in Silos to illustrate this concept. After harvest, the grains need to be transported from the field, and in fact, material handling begins right at the harvest stage, as you will see later. These grains must then be moved to a facility where they will be cleaned and processed before being stored in the Silo. Since Silos are often tall structures, material handling equipment plays a crucial role in ensuring the safe and efficient transfer of the grains into these storage units. And when the grains need to be removed from the Silo, the same equipment is once again employed to facilitate the movement. This clearly demonstrates the pivotal role of material handling in the post-harvest phase of agriculture.

As with most processes today, material handling can vary in terms of sophistication. Let's quickly go through these different levels:

- **Manual Material Handling:** While this may seem obvious, it's important to mention that this method relies entirely on human labor for tasks such as lifting, packing, and loading. However, this approach comes with its fair share of risks, as much of the work involves lifting with the back, which can cause strain and injury. To address this, various ergonomic tools like hoists and lift tables have been developed to reduce the physical strain of manual handling. The primary reason for these improvements is to protect workers from the dangers associated with handling materials manually.

Figure: 5.2 Robots at Work

- **Semi-automated Material Handling:** This method is employed by farms looking to minimize the risk to workers but may not have the resources to fully automate their material handling processes. In semi-automated systems, a machine handles a specific task, but a human, such as a driver or operator, is responsible for controlling or overseeing the operation. This is a common approach on many farms, though it is often confused with another method of handling, which we'll discuss later.
- **Automated Material Handling:** This is the fully automated approach, where sophisticated technology and machines replace manual labor for handling materials. In this system, robots and other advanced technologies carry out all the tasks involved in material handling. Some farms have already begun using robots as early as the harvest stage, marking a significant shift towards automation in agriculture.

This method is particularly useful when the materials being handled require the highest hygiene standards and zero tolerance for contamination.

These farms use sensors on the animals, which detect when the milk volume is at the optimal level for milking. Once this is identified, the sensor triggers the animal to move toward the milking stand. Once there, a machine performs the milking, ensuring the entire process is completed without any human intervention.

Integrated Material Handling: As discussed, automated material handling is often confused with semi-automated handling, so let me clarify the difference using a fun analogy. Picture yourself as a transporter responsible for moving celebrities between two cities. Most of the time, you drive them halfway to their destination, but once you reach a certain point, you have them exit the car and walk along a red carpet, where beautiful ladies usher them into a driverless car that you've programmed to complete the rest of their journey on its own.

In this scenario, your initial driving would represent semi-automated handling, as you're still in control, but the actual transportation process is partly automated. The ladies ushering the celebrities are performing manual handling, and finally, the driverless car would symbolize automated handling, as it operates entirely on its own.

Now, if your company were to combine all these methods—manual handling (the ushers), semi-automated handling (your driving), and fully automated handling (the driverless car)—it would be considered an Integrated Transporting Company. For a process to be deemed "integrated," it must combine at least two of the previously mentioned handling methods.

Agricultural Material Handling Equipment

Now, let me quickly introduce you to three main divisions of material handling equipment commonly used in agriculture. While there are other types, these three are the most prevalent on many farms. Let's take a closer look at Conveyors, Elevators, and Cranes/Trucks.

- **Conveyors:** They are mechanical tools used for moving bulk materials from one place to another. Now let's discuss three types of conveyors:

Figure: 5.3 Screw Conveyor

- **Screw Conveyors:** If you chose to call it an Auger Conveyor, you're still very correct. These basically move materials by employing the principle of rotation. A typical one has a helical screw that is usually mounted on a shaft. This shaft when driven by electric motors or tractors (through power transmission shafts) moves materials ranging from cereals to animal feed at a rate that is proportional to the speed of rotation and the angle of inclination in inclined types. These screws can either be positioned horizontally or at an incline. If you're thinking that you've never come across these screws before, it's likely because they're typically housed within tubes or troughs. However, if you've observed how a combine harvester moves harvested grains into trucks, you've seen this type of screw in action. If you haven't, this image should give you an idea for the time being.

- **Belt Conveyors :** These conveyors are also built on the principle of rotation albeit they do so it a arrangement of a belt and two pulleys; a powered pulley (connected to the power source) and the idler pulley. The belts are mounted on these pulleys, and the materials to be moved are placed on the belts, such that as the pulleys rotate the belts, the materials are transported. These materials range from fruits to grains and even boxes of packaged products.

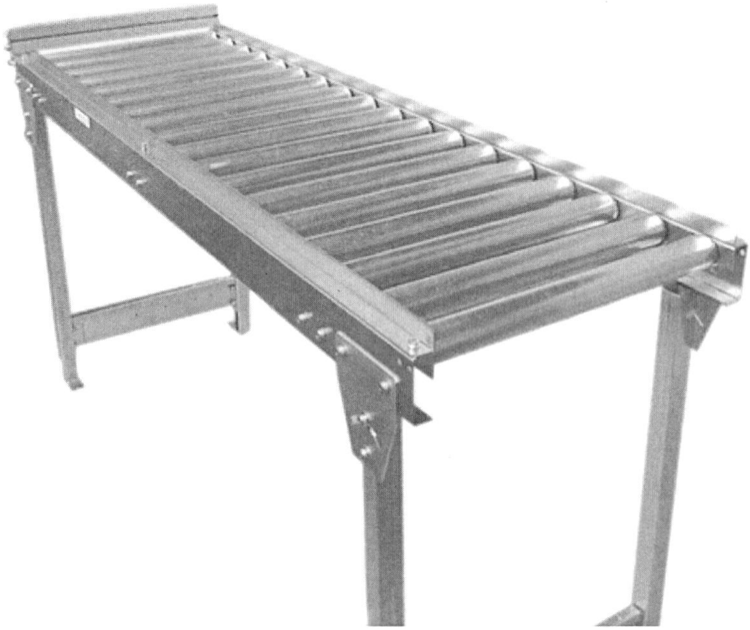

Figure: 5.4 A Roller Conveyor

- **Roller Conveyors:** These conveyors are used for carrying bulk materials that are larger as those carried by the previous conveyors. They employ rollers that are driven by a major shaft. The shaft and rollers are connected by spots. Usually, the shafts are driven by electric motors. The fact that the rollers are connected to the shafts by spools are said to have a safety benefit, for slip and for trappings in the system.
- **Bucket Elevators:** This is the type of elevators mainly used in agriculture. Usually an arrangement of buckets and belts, or sometimes chains. The bucket scoops the material from a pile at a lower level and holds it parallel to the floor until it reaches a higher point where it deposits the material by tipping the bucket. There are three major types: Centrifugal Discharge Elevators- mainly used for grains in grain elevators. Continuous Discharge Elevators and Positive Discharge Elevators which are used to convey completely processed food materials like popcorn.

Figure: 5.5 A Crane by Siddharthnarain

- **Cranes/Trucks :** These two are together because they usually work side by side in Agricultural Material handling. Cranes are usually used for unloading packed raw materials or products from trucks, for stacking them for storage before or after processing in warehouses and for loading packed final products into trucks.

Principles of Material Handling

The principles of material handling are as follows:

- **Orientation Principle:** This principle emphasizes the importance of thoroughly studying all relevant system relationships before initiating preliminary planning. It involves reviewing existing methods, identifying problems, and analyzing current practices.
- **Planning Principle:** It advocates for the creation of a comprehensive plan that includes basic requirements, potential alternatives, and contingency measures.
- **Systems Principle:** This principle integrates handling and storage activities into a cost-effective system design that ensures efficiency across all processes.
- **Unit Load Principle:** The goal is to handle products in the largest possible unit load to maximize efficiency.
- **Space Utilization Principle:** It stresses the effective use of all available space, optimizing storage and movement within a given area.
- **Standardization Principle:** This principle encourages the standardization of handling methods and equipment to improve consistency and reduce variation.

- **Ergonomic Principle:** It considers human capabilities and limitations by designing handling equipment that minimizes physical strain and enhances worker safety.
- **Energy Principle:** The energy consumed during material handling is a critical consideration, aiming to optimize energy use and reduce waste.
- **Ecology Principle:** This principle advocates for minimizing environmental impact during the material handling process, ensuring sustainability.
- **Mechanization Principle:** It promotes the use of mechanized handling processes whenever possible to increase efficiency and reduce human labor.
- **Flexibility Principle:** This principle encourages the use of methods and equipment that can adapt to various conditions and changing requirements.
- **Simplification Principle:** Simplification of processes and methods by eliminating unnecessary movements is key to improving efficiency.
- **Gravity Principle:** This principle suggests utilizing gravity in the movement of goods to reduce energy consumption and enhance productivity.
- **Safety Principle:** Safety should be prioritized, ensuring that handling equipment complies with safety regulations and minimizes risk to workers.
- **Computerization Principle:** It encourages the use of computerized systems to manage material handling and storage, improving accuracy and streamlining processes.
- **System Flow Principle:** This principle advocates for the integration of data flow with the physical movement of materials, ensuring alignment and reducing inefficiencies.
- **Layout Principle:** A detailed operational sequence of all systems available should be prepared to ensure smooth and effective flow.
- **Cost Principle:** The principle encourages conducting a cost-benefit analysis of all possible solutions to ensure economic efficiency.
- **Maintenance Principle:** It emphasizes the importance of planning for preventive maintenance and regular repairs to ensure continuous system performance.
- **Obsolescence Principle:** This principle advocates for the creation of an equipment policy that ensures the use of up-to-date technology and maximizes economic benefits.

Material handling operations are designed based on the principles outlined above. The equipment used in these operations typically includes cranes, conveyors, and industrial trucks.

6

Handling, Packing and Storage of Agricultural Commodities

The handling, packing, and storage of agricultural commodities are key elements of the post-harvest process that have a direct impact on the quality, safety, and shelf life of food products. Following harvest, various agricultural products like grains, fruits, vegetables, and other perishables undergo a series of careful steps to maintain their optimal condition until they reach either the consumer or the processing phase. Efficient handling reduces damage, contamination, and waste, while proper packaging acts as a protective barrier, safeguarding products from environmental factors and making transportation and sales smoother. The storage phase is important in maintaining the quality of the commodities, preventing spoilage, and preserving their nutritional content. critical factors such as temperature regulation, humidity control, pest management, and the use of the right packaging materials all play a significant role in prolonging the shelf life of agricultural products. In summary, the efficiency and effectiveness of the handling, packing, and storage systems have a substantial influence on the economic sustainability of the agricultural industry, help minimize waste, and ensure food security for consumers.

TYPES AND USES OF PACKING AND PACKAGING MATERIALS

Packaging plays a crucial role in the handling system, acting both as an asset and a potential obstacle to maintaining the maximum shelf life and quality of agricultural products. The appropriate packaging needs to allow for proper ventilation while also being strong enough to prevent collapse. When produce is packed for easy handling, containers like waxed cartons, wooden crates, or rigid plastic containers are preferable to bags or open baskets. This is because bags and baskets offer no protection when the products are stacked. Sometimes, locally made containers can be reinforced or lined to provide extra protection. While waxed cartons, wooden crates, and plastic containers may be more costly, they are still a cost-effective solution for domestic markets due to their reusability and ability to withstand the high humidity often found in storage conditions. Adding simple cardboard liners to crates aids in reducing abrasion to the produce.

For optimal results, containers should neither be packed too loosely nor too tightly. Loose products can vibrate against one another, causing bruising, while over-packing can result in compression bruising. Shredded newspaper serves as an inexpensive, lightweight filler for shipping containers Various agricultural fibers are suitable for paper production, and handlers may find it economically practical to incorporate these operations into their post-harvest systems. Corrugated fiberboard is available in four types of flutes, with type B (1/8 inch in height, 47 to 53 flutes per inch, and a basis weight of 26 lb per 1,000 ft^2) being the most commonly used for perishable goods handling.

In high-humidity environments, packaging can lose much of its strength, leading to collapsed packages that offer little to no protection, forcing the commodity inside to bear the weight of the load above. While packing is meant to protect goods by immobilizing and cushioning them, it can make temperature management more challenging if it blocks ventilation. Packing materials can also act as vapor barriers, helping maintain higher humidity levels within the package. Besides protection, packaging allows for quicker handling throughout the distribution and marketing process, minimizing the impact of rough handling.

Produce can be hand-packed to create visually appealing packages, often using a fixed count of uniformly sized units. Packaging materials such as trays, cups, wraps, liners, and pads are often added to help secure the produce. Basic mechanical packing systems typically use volume-fill or tight-fill methods, in which sorted produce is placed into boxes and then vibration-settled. Many volume-fillers use weight as an estimate of volume, with final adjustments made by hand.

Ethylene absorber sachets, placed inside containers with ethylene-sensitive produce, can slow down the ripening of fruits, the de-greening of vegetables, and floral wilting. Packaging using plastic films can modify the atmosphere around the produce, a technique known as modified atmosphere packaging (MAP). MAP typically restricts air movement, allowing the product's natural respiration process to reduce oxygen levels and increase carbon dioxide levels inside the package. A significant benefit of plastic films is the reduction of water loss.

MAP can be applied within both shipping containers and consumer units. Atmospheric modifications can be actively created by generating a slight vacuum in a vapor-sealed package (like an unvented polyethylene bag) and replacing the internal atmosphere with a desired gas mixture. In most cases, lowering oxygen levels and increasing carbon dioxide will benefit the majority of commodities. The choice of the best polymeric film for each commodity/package size combination depends on the film's permeability and the product's respiration rate under expected time/temperature conditions during handling. Oxygen, carbon dioxide, and/or ethylene absorbers can be incorporated into packages or containers to maintain the desired atmospheric balance.

Modified atmosphere packaging should always complement proper temperature and humidity control. The difference between beneficial and harmful concentrations of oxygen and carbon dioxide is relatively small for most products, so caution must be exercised when utilizing these technologies.

Best Practices in Packing

The packing stand shown in the diagram can be securely bolted to another stand of the same design if additional space is needed for packing produce. If trimming is required, a loose board should be added, ensuring it is thick enough to extend above the height of the front rail. It's important for the front rail to be smooth and rounded for ease of use.

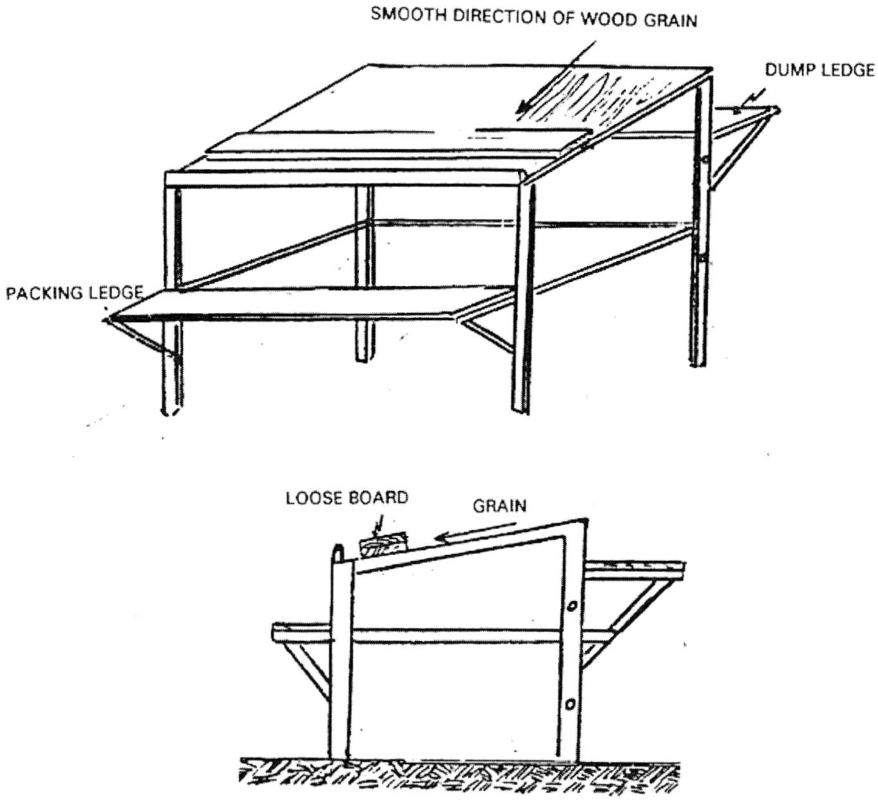

Figure: 6.1 The Packing Stand

A basic field packing station can be built using wooden poles and a sheet of polyethylene. A thatched roof will offer shade and help maintain a cool environment. The structure should be positioned in such a way that the roof overhang effectively blocks most of the sun's rays.

Hands of bananas, after undergoing washing to remove latex and perhaps spraying with fungicides, are typically packed into cardboard containers lined with polyethylene. The following illustrations depict one method of filling a container with the fruit in order to ensure less damage during transport. Note that the polyethylene liner is folded up over the bananas before closing the box.

Figure: 6.2 Wide, Flat Medium-to-Small Hand in Middle of Compartment

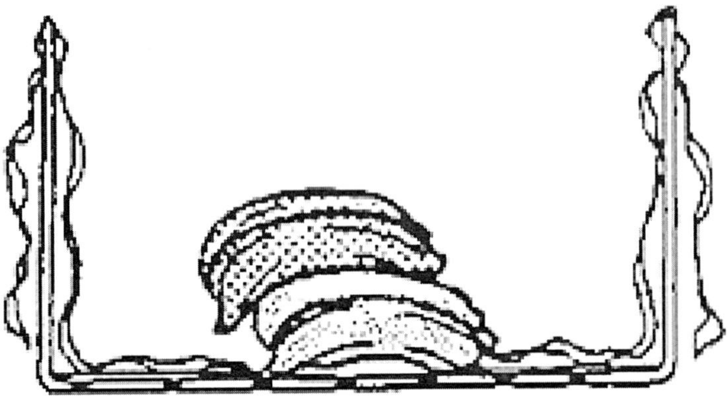

Figure: 6.3 Medium-length, Wide Hand on top, Crown not Touching Fruit Below

Figure: 6.4 Medium-to-Short Length, Wide Hand, Crown not Touching Fruit Below

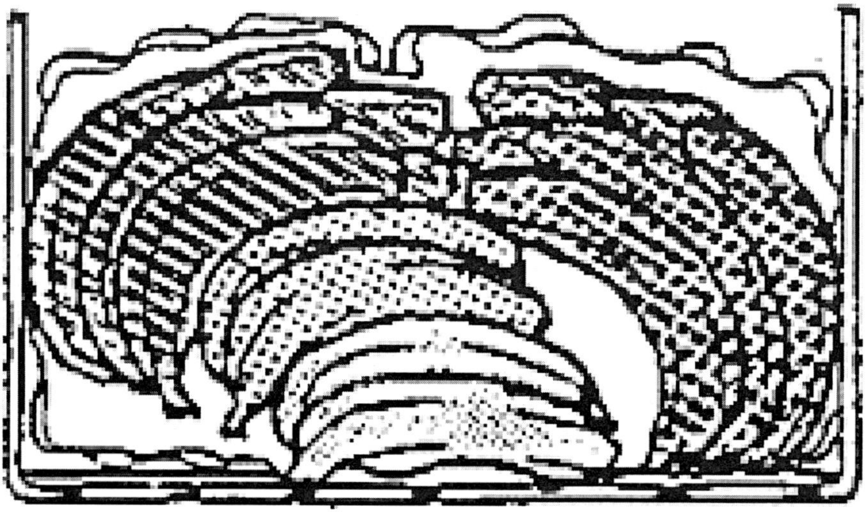

Figure: 6.5 One Large Hand, or Two Clusters With Long Fingers

A circular rotating table can be used to pack a variety of crops. The produce is fed in along a conveyor or if no conveyor is in use, simply put onto the table, where packers select the produce and fill cartons at their stations. In the illustration below, a discard belt has been added below the supply belt, allowing easy disposal of culls.

Each packer can work independently, trimming as needed and check weighing cartons on occasion.

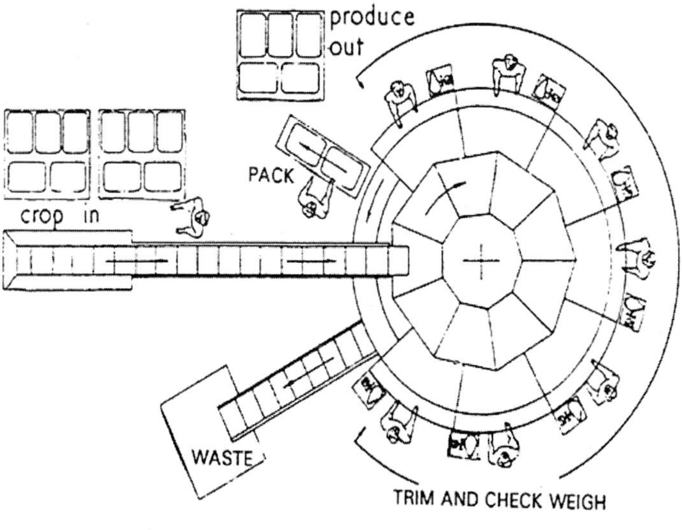

Figure: 6.6 A Circular Rotating Table to Pack the Crops

Types of Packing Containers

Sacks are usually used for packaging produce because they are cost-effective and easily accessible. However, the table below outlines the characteristics of various materials used to make sacks. It is important to note that none of these sack types offer adequate protection for fresh produce and should be avoided whenever possible.

Characteristics of Sacks as Packaging Units

Sack types	Tearing and snagging	Impact	Protection against		Contamination	Notes
			Moisture absorption	Insect invasion		
Jute	Good	Good	None	None	Poor, also cause contamination by sack fibers	Bio-deterioration. Insect harborage. Odor retention.
Cotton	Fair	Fair	None	None	Fair	High re-use value.
Woven plastics	Fair-Good	Good	None	Some protection (if closely woven)	Fair	Badly affected by ultra-violet light. Difficult to stitch.
Paper	Poor	Fair-Poor	Good - WFP multi-wall sacks have plastic liners.	Some protection, better if treated.	Good	Consistent quality. Good print.

The table below provides examples of some types of typical mechanical damage and their effect on packaging containers.

Type of damage	Container	Result	Important factors
Impact damage through dropping	Sacks - woven and paper	Splitting of seams and material causing leaking and spillage loss.	Seam strength
	Fiberboard boxes	Splitting of seams, opening of flaps causing loss of containment function. Distortion of shape reducing stacking ability.	Bursting strength Closure method
	Wooden cases	Fracture of joints, loss of containment function.	Fastenings Wood toughness
	Cans and drums	Denting, rim damage. Splitting of seams and closures causing loss of containment and spoilage of contents.	
	Plastic bottles	Splitting or shattering causes loss of contents.	Material grade Wall thickness

Compression damage through high stacking	Fiberboard boxes	Distortion of shape, seam splitting causing loss of containment and splitting of inner cartons, bags, and foil wrappings.	Box compression strength
	Plastic bottles	Distortion, collapse and sometimes splitting, causing loss of contents.	Design, material, wall thickness
Vibration	Corrugated fiberboard cases	Become compressed and lose their cushioning qualities. Contents more prone to impact damage.	Box compression strength
Snagging, tearing, hook damage	Sacks - woven and paper	Loss of containment function - spillage (more severe with paper sacks).	Tear strength
	Tins	Punctured, loss of contents.	Metal thickness

The diagrams below are for a variety of commonly used fiberboard containers. Final dimensions can be altered to suit the needs of the handler, and all containers should have adequate vents. Cartons can be glued, taped or stapled as desired during construction at the packinghouse.

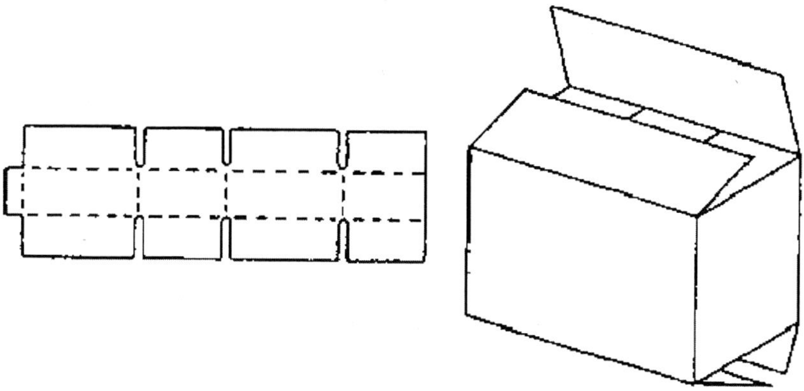

Figure: 6.7 One Piece Box

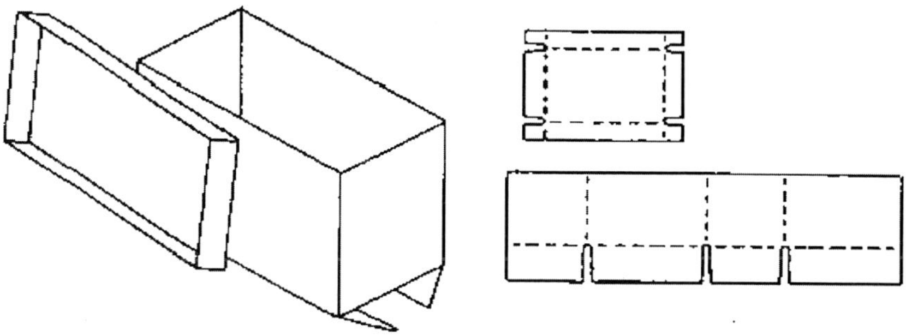

Figure: 6.8 Two-Piece Box With Cover

The diagrams below are for a variety of commonly used fiberboard containers. Final dimensions can be altered to suit the needs of the handler, and all containers should have adequate vents.

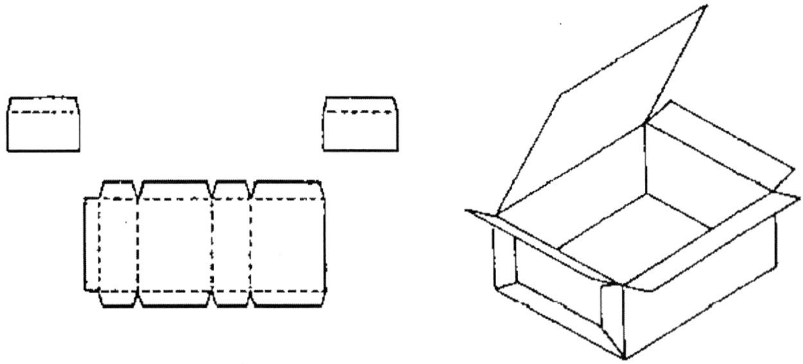

Figure: 6.9 Bliss-Style Box

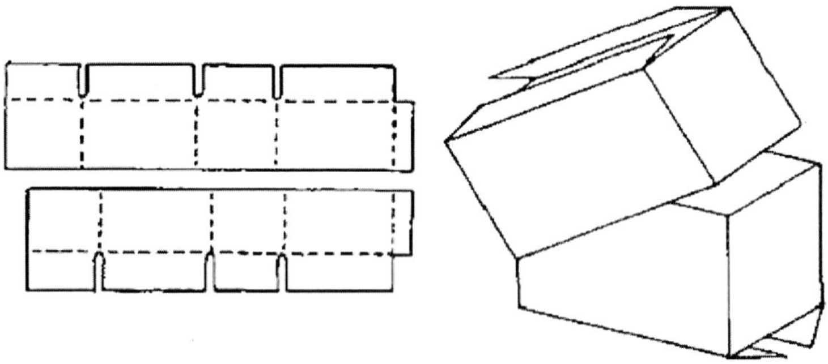

Figure: 6.10 Full Telescoping Box

The diagrams below are for a variety of commonly used fiberboard containers. Final dimensions can be altered to suit the needs of the handler, and all containers should have adequate vents.

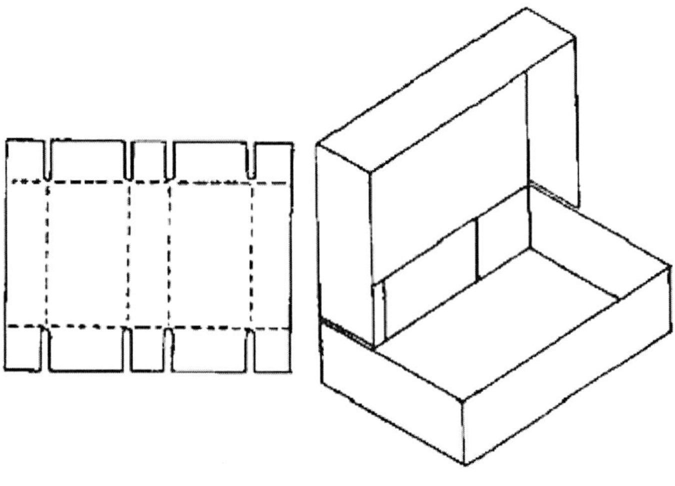

Figure: 6.11 One-Piece Telescoping Box

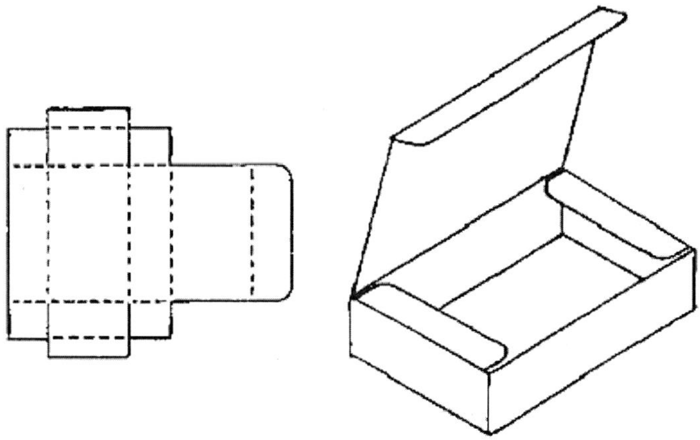

Figure: 6.12 One-Piece Tuck-in Cover Box

The diagrams below depict a variety of commonly used fiberboard containers. The final dimensions of these containers can be adjusted to meet the specific requirements of the handler. It's important to ensure that all containers have sufficient ventilation to maintain the quality of the contents.

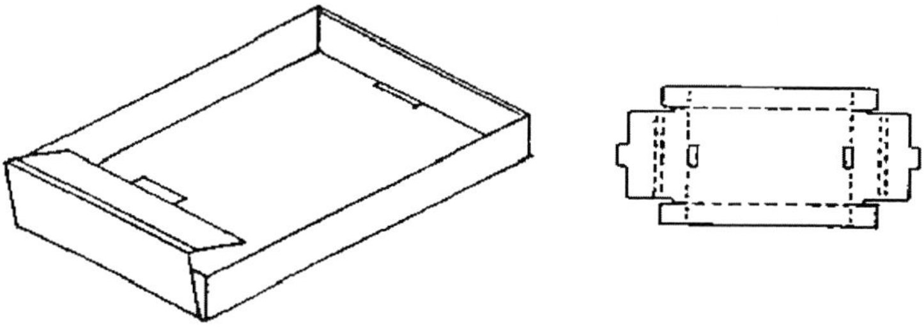

Figure: 6.13 Self-Locking Tray

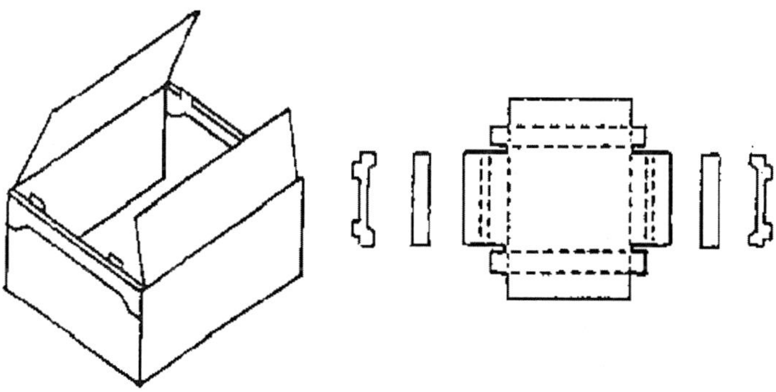

Figure: 6.14 Interlocking Box

Containers can be constructed from wood and wire, using the general diagrams provided below. A special closing tool makes bending the wire loops on the crate's lid easier for packers to do. Wire-bound crates are used for many commodities including melons, beans, eggplant, greens, peppers, squash and citrus fruits. Package Research Laboratory can provide a list of suppliers in your area.

Figure: 6.15 General Diagram of Container Constructed from Wood and Wire

Here is a simple illustration of the parts of a wooden crate.

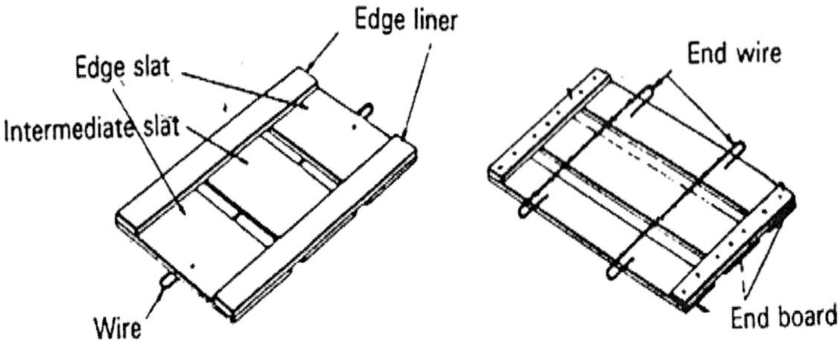

Figure: 6.16 Parts of a Wooden Crate

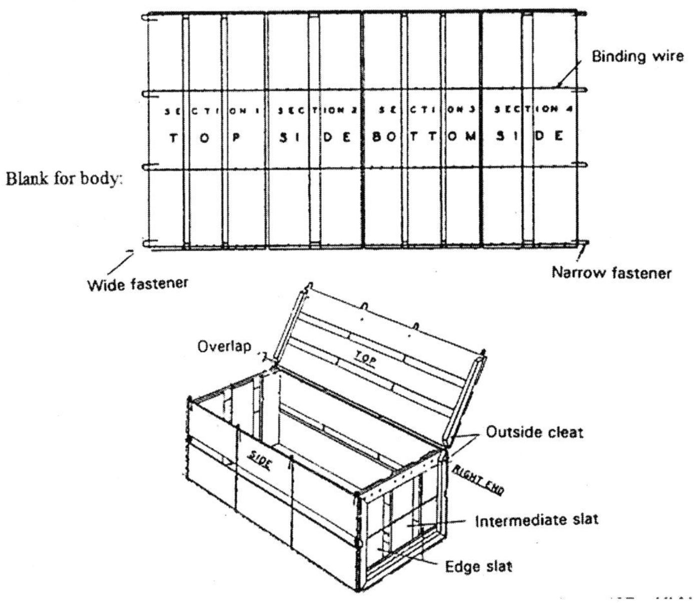

Figure: 6.17 Blank for Body

A wooden lug is the standard packing container for table grapes, known for its durability and ability to maintain its stacking strength even under high humidity over extended periods. Rigid plastic containers are also commonly used.

Before sealing the container, a paper liner is often folded over the grapes to protect them from dust and moisture buildup. Additionally, a pad containing sulfur dioxide may be enclosed with the grapes inside a plastic liner to help control decay. However, it's important to note that sulfur dioxide treatments can cause damage (such as bleaching) to most other commodities.

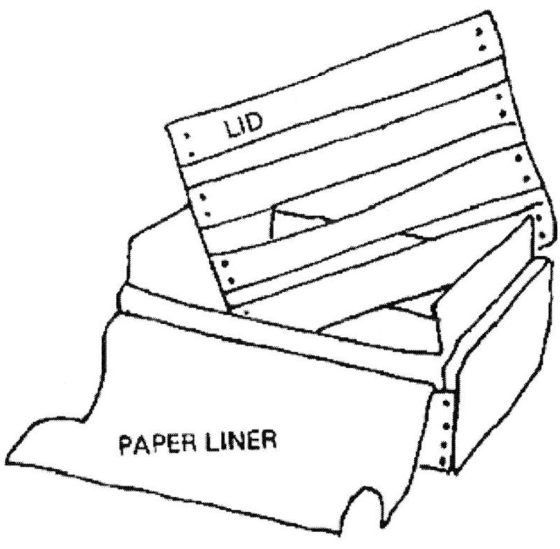

Figure: 6.18 Paper Liner

Rigid plastic or wooden containers are also used extensively for asparagus. The trimmed spears are packed upright in containers that provide for a large amount of ventilation.

Figure: 6.19 A Rigid Plastic Container

For domestic marketing, plastic crates provide excellent protection for produce and adequate ventilation during handling, cooling, transport and storage. Some plastic crates are collapsible or can be nested when stacked for easier handling when empty. Crates should be cleaned on a regular basis with chlorinated water and detergent to reduce the chances of spreading decay from one load to the next.

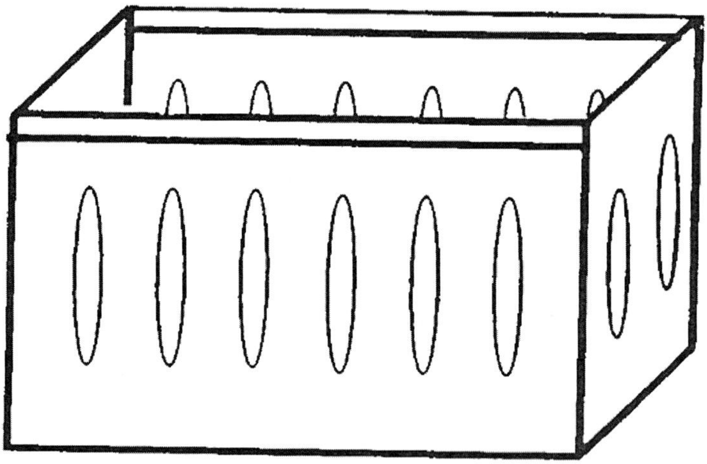

Figure: 6.20 A Plastic Crate Provides Protection to Produce and Ventilation During Handling, Cooling, Transport and Storage

Containers for cut flowers are typically designed to be long and narrow with a full telescopic shape, featuring vents at both ends to support forced-air cooling. The total vent area should make up about 5% of the overall surface area of the box. To help maintain cool temperatures, especially when boxes are temporarily delayed in transport or storage in environments with uncontrolled temperatures, a closable flap can be added to the container.

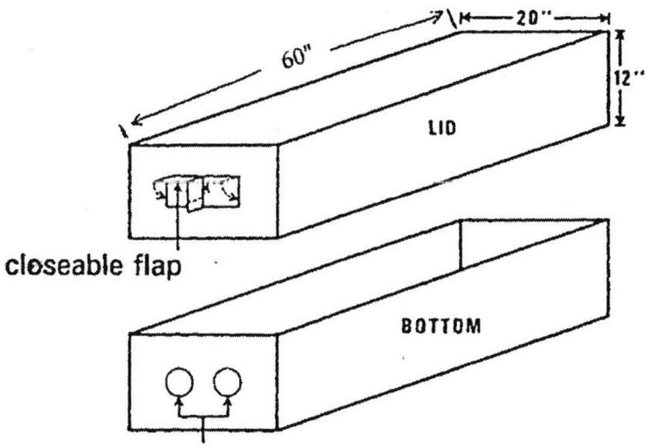

Figure: 6.21 Containers for Cut Flowers

Source: Rij, R. et al. 1979. Handling, precooling and temperature management of cut flower crops for truck transportation. USDA Science and Education Administration, AAT-W-5, 26 pp (Issued jointly as University of California Division of Agricultural Sciences Leaflet 21058).

A simple wooden tray with raised corners is stackable and allows plenty of ventilation for fragile crops such as ripe tomatoes.

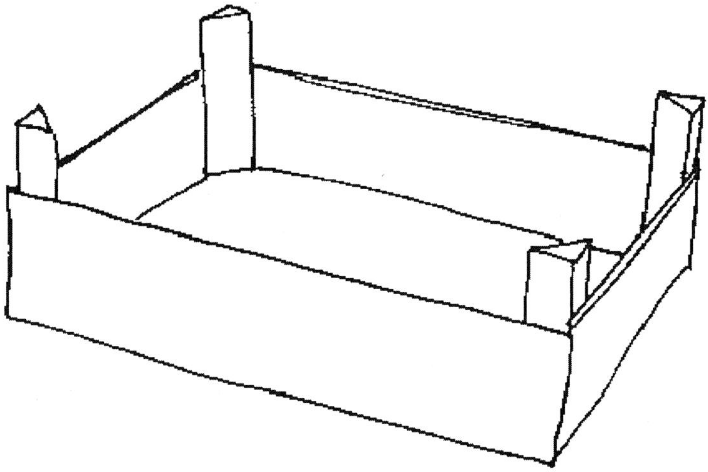

Figure: 6.22 A Wooden Tray With Raised Corner

KEY INFLUENCES ON STORAGE CONDITIONS

The storage conditions of agricultural products can be influenced by various factors, including environmental and product-related factors.

Environmental factors, including temperature, humidity, light, and air quality, play a crucial role in determining the shelf life and overall quality of crops. For example, high temperatures can accelerate respiration and water loss, leading to decreased quality and faster spoilage. Excessive humidity, on the other hand, can promote mold and bacterial growth, while exposure to light can cause nutrient degradation and diminish color and flavor.

Product factors, such as the type of crop, its maturity level, packaging materials, and moisture content, also significantly impact storage conditions. Different crops require different storage conditions, and understanding these requirements is essential to maintaining their quality and market value. For instance, fruits and vegetables need specific storage conditions based on their ripeness, with some requiring cooler temperatures and lower humidity levels than others. The choice of packaging materials, such as plastic, paper, or wax, can either hinder or encourage moisture loss, affecting storage effectiveness.

Moisture content is another critical factor. Excess moisture can lead to mold and bacterial growth, while insufficient moisture may result in product dehydration. It's vital for farmers and food distributors to account for these factors and implement the most suitable storage practices to prevent spoilage and preserve the quality of their agricultural products.

Optimal Storage Conditions for Specific Crops

Each type of crop has unique ideal storage conditions that are necessary to preserve its freshness and quality. For example, fruits generally thrive in a cool, dry environment with low humidity to prevent spoilage. However, certain fruits such as apples and pears need colder temperatures than others like citrus fruits. Similarly, vegetables require specific storage conditions depending on their ripeness, with some needing cooler temperatures and lower humidity to remain fresh. Grains and nuts should be stored in a cool, dry area to avoid moisture-related damage and insect infestations. Dairy and meat products, on the other hand, must be kept in the refrigerator to prevent spoilage and the growth of harmful bacteria. Understanding the proper storage conditions for each type of crop is vital for maintaining their quality and market value.

Effective Packaging and Handling

Ensuring proper packaging and handling is key to extending the shelf life and maintaining the quality of agricultural products. The right packaging protects products from various types of damage, including physical, chemical, and biological harm during storage and transport. It also helps regulate moisture, prevent insect infestations, and reduce spoilage. The type of packaging material, such as plastic, paper, or wax, depends on the product, the storage environment, and the desired shelf life. Alongside packaging, appropriate handling procedures are essential to avoid damaging the products. Farmers and food distributors should handle the produce carefully to avoid bruising, cutting, or puncturing, which can lead to spoilage.

Furthermore, maintaining the correct temperature and humidity levels during storage is crucial for preventing any damage. By following proper packaging and handling practices, agricultural products will stay fresh and high-quality for extended periods, reducing waste and boosting profits.

Storage Conditions Monitring and Maintenance

Ongoing monitoring of storage conditions is essential for maintaining the quality of agricultural products. Farmers and food distributors should use monitoring devices like thermometers, hygrometers, and CO_2 sensors to measure temperature, humidity, and air quality. In the event of deviations from optimal conditions, corrective actions should be taken immediately to prevent spoilage. Maintenance of storage equipment is just as important to ensure that these monitoring systems remain accurate and reliable. Regular calibration, cleaning of equipment, and upkeep of storage facilities are crucial for creating an optimal storage environment. By properly monitoring and maintaining storage conditions, agricultural products can be stored in the best conditions to ensure their maximum shelf life and quality.

Safety Considerations

Storage facilities for agricultural products can present several safety risks, including fire hazards, chemical spills, and the potential for structural failure. To mitigate these risks, farmers and food distributors should conduct regular safety inspections, provide safety training for employees, and ensure that hazardous materials are stored properly. Emergency procedures should also be established to deal with safety hazards, ensuring that employees and products are protected in case of an accident. By focusing on safety in agricultural storage, the risk of accidents is minimized, ensuring the safety of workers and the integrity of the products.

Compliance with Regulatory Standard

Compliance with these regulations is essential for ensuring that agricultural products are safe and of high quality, as well as to avoid public health risks. Farmers and food distributors should stay informed about these regulations and ensure they meet the necessary requirements, which may include implementing proper storage practices and maintaining detailed records of storage conditions.

Classification of Material Management

Material management experts typically focus on five distinct areas, each of which plays a vital role in optimizing material use and production efficiency:

Strategic Planning for Material Requirements

This crucial aspect of material management has a direct impact on a company's profits. The less material used, the lower the production costs, resulting in higher profit margins. Reducing material wastage has led some industries to adopt "Just in Time" (JIT) strategies, which involve keeping inventory levels minimal. However, even with JIT, careful planning is essential to ensure production schedules are not affected, and material needs are met without overstocking.

Purchasing

Purchasing should be done in a cost-effective and timely manner to ensure the continuous availability of materials while minimizing expenses. Timely purchases not only maintain material supplies but also help to maximize profits by controlling costs and ensuring a steady flow of materials needed for production.

Inventory Control

Inventory management includes monitoring a range of materials, from partially finished goods to items ready for sale or use in production. Efficient inventory control ensures materials are available when needed without overstocking. Many industries align their purchasing schedules so that materials arrive just in time for production, while also maintaining sufficient stock levels to avoid stockouts. Proper inventory management helps control the flow of raw materials, finished goods, and parts essential for continued operations.

Management Material Supply

Managing the supply of materials across multiple sites or production centers is a key aspect of supply chain management. Any disruptions in material supply, whether due to stock shortages or damage, can lead to costly delays and halt production processes. Efficient material supply management involves organizing stock levels, ensuring timely delivery of materials, and using backup systems to prevent delays or shortages that could affect production.

Quality Control

Quality control is critical in materials management, as the quality of raw materials directly influences the final product. Key quality factors include durability, dimensional accuracy, performance, reliability, and appearance, all of which are important depending on the specific application. Ensuring that materials meet high-quality standards results in superior products and minimizes the likelihood of defects or returns.

All the above areas of material management must work cohesively to effectively manage materials from the initial purchase stage through to their use in production.

Role of Material Managers

The primary goal of material managers is to ensure a constant and efficient flow of materials throughout the production process. Although this may appear to be a simple task, it involves overcoming several challenges, such as inaccurate bills of materials, incorrect stock counts, errors in shipping and receiving, unaccounted-for scrap, and issues with production reporting.

Effective planning, organizing, and controlling the movement of materials allow for better coordination of purchasing and shipping with the manufacturing process and the final product delivery. While material managers are responsible for overseeing the company's inventory management, the actual procurement of materials might be handled by a separate purchasing team.

In addition to ensuring that materials are delivered on time, material managers also focus on managing costs and maintaining quality throughout the supply chain. By monitoring the availability of raw materials and products, they can help achieve cost savings and ensure the highest return on working capital.

Materials are typically divided into two categories: direct and indirect materials. Direct materials are those that are essential to producing a finished product, while indirect materials are those that do not directly contribute to the final product's creation.

Regardless of the category, inventory management plays a critical role in material management. This can be broken down into the following key components:

Maximum Stock

This refers to the largest quantity of material that can be stored at any given time, ensuring that there is enough to meet production demands.

Minimum Security Stock

Given the fluctuations in stock levels during production, it's essential to establish a minimum stock threshold, considering factors like supplier lead times, order costs, and production needs.

Re-Order Point

This is the inventory level at which new orders should be placed to ensure that warehouse supplies remain in line with supplier delivery schedules and production timelines.

7

Transportation and Marketing of Agricultural Commodities

The transportation and marketing of agricultural products play critical roles in the supply chain, ensuring that farm goods reach consumers effectively and at competitive prices. Once agricultural items are harvested, they must be transported from farms to processing plants, wholesalers, and retail stores, which often requires covering long distances using various modes of transportation, including trucks, trains, and ships. The transportation process must be meticulously managed to prevent damage, spoilage, or loss, particularly for perishable items such as fruits, vegetables, and dairy. Equally essential is the marketing of these products, which involves implementing strategies to promote goods, generate demand, and link producers with buyers.

Effective marketing encompasses branding, packaging, pricing, and distribution through diverse channels like local markets, supermarkets, and export outlets. Both transportation and marketing not only affect the final cost of agricultural goods but also influence food security, the efficiency of the supply chain, and the economic well-being of farming communities. A well-organized transportation and marketing system ensures that agricultural commodities are delivered promptly and affordably while satisfying consumer needs for quality and variety.

MARKETING STRATEGIES FOR AGRICULTURAL PRODUCE

Agricultural marketing is a decision-making process that begins with the choice to produce a marketable farm product. It encompasses all the institutional and functional aspects of the market structure within the system, influenced by both technical and financial factors.

- It is the primary method for transporting agricultural goods from farms to markets and various urban areas.
- The agricultural marketing process involves acquiring, storing, preparing, shipping, and distributing a wide range of farm products nationwide.

- Agricultural marketing, as part of the broader marketing system, refers to the entire range of activities, organizations, and regulations involved in acquiring farm inputs and moving agricultural products from farms to consumers, manufacturers, and exporters.
- It aims to provide consumers with food that meets acceptable quality standards at affordable prices, ensure fair prices for farmers, and offer adequate margins for intermediaries to maintain their ongoing role in the trade.
- The success of the agricultural industry as a whole is dependent on efficient agricultural marketing.
- Fair compensation for farmers' agricultural products ensures their financial stability.
- Furthermore, it ensures that food products are available and affordable for consumers, helping to curb inflation.
- The flow of agricultural commodities from producers to consumers is managed by a comprehensive system called agricultural marketing.
- In India, agricultural marketing is facilitated through various central government agencies, such as the Commission of Agricultural Costs and Prices, the Food Corporation of India, the Cotton Corporation of India, and the Jute Corporation of India, among others.
- There are also specialized organizations responsible for the marketing of products like rubber, tea, coffee, tobacco, spices, and vegetables.

Significance of Transport and Marketing on Agricultural Produce

Agriculture and transportation are two interconnected sectors of the economy, heavily reliant on each other. The effectiveness of transportation infrastructure plays a significant role in the marketing of agricultural products.

- Unlike industrial goods, agricultural products have unique characteristics that make transportation quality and availability essential.
- For example, agricultural products tend to be bulky and perishable, with almost all of them being consumable goods.
- Proper packaging and careful transportation are required to ensure that these goods remain undamaged during transit.
- When efficient transportation infrastructure is in place, farmers are better able to maximize their profits while ensuring their agricultural output is handled with the necessary care.
- An effective transportation and marketing system helps maintain low unit prices and strengthens the agricultural value chain.
- Affordable transportation enables producers to achieve profitability while ensuring their products are accessible to customers.
- Conversely, high transportation costs can negatively affect local marketing and hinder the ability to export agricultural products to countries with more efficient transportation systems.

- Transportation plays a vital role in creating markets for agricultural goods, improving communication between different regions, and opening up new economic opportunities.
- Investing Rs 89,375 crore in the development of efficient post-harvest storage and transportation could generate over 3 million jobs.
- Poor rural transportation quality can negatively impact productivity, income, and the overall standard of living.
- The primary goal of agricultural marketing is to facilitate the buying and selling of agricultural goods.
- The ability to sell agricultural products depends on various factors, including current demand and the availability of storage space.
- Improved marketing strategies can ensure higher prices for farmers' produce while protecting them from price fluctuations.
- Manufacturers aim to receive affordable, high-quality produce from farmers to ensure profitability in their sales.
- Traders and retailers seek to acquire the best products and services from manufacturers or farmers at competitive prices.
- Consumers want high-quality goods at reasonable prices.
- An effective marketing strategy can reduce the need for government purchases at Minimum Support Prices (MSP), helping to contribute to fiscal consolidation.

SCOPE OF AGRICULTURE MARKETING

Given the size and diversity of India's agricultural sector, the potential for agriculture marketing there is enormous. The agriculture industry is a major contributor to India's economic growth and development, as over 60% of the country's population makes their living from it. The following are some of India's main agricultural marketing opportunities:

1. Value chain expansion: For farmers and other agricultural stakeholders, the creation of effective value chains connecting producers, processors, and markets may open up new revenue streams. This calls for training, technology, and infrastructure investments in addition to productive collaborations amongst various value chain participants.
2. Adoption of contemporary marketing techniques: Farmers can increase their market access and price realization by utilizing contemporary marketing techniques like contract farming, supermarkets, and e-commerce platforms. These techniques can also boost the efficiency and competitiveness of the agriculture industry and encourage the adoption of contemporary farming methods.
3. Encouragement of direct marketing: Farmers can offer better prices and a closer relationship with their clients through direct marketing platforms like farmers' markets and community-supported agriculture (CSA). Since customers can track the provenance and caliber of the goods they buy, these channels can also encourage accountability and transparency in the agriculture industry.

4. Growth of rural markets: India's rural markets are a significant source of demand for agricultural goods, and their expansion may open up new markets for farmers. To increase the effectiveness and competitiveness of rural markets, this calls for investments in information systems, transportation, and market infrastructure.
5. Encouragement of export-oriented agriculture: India has the potential to grow into a significant exporter of agricultural goods, especially in sectors like spices, horticulture, and floriculture. Effective collaborations between producers, exporters, and government organizations are necessary for the growth of export-oriented agriculture, as are investments in marketing, logistics, and quality control.
6. Technology use: India's agricultural marketing can become more transparent and efficient by utilizing technologies like blockchain, precision agriculture, and mobile applications. In the agricultural industry, these technologies can support fair trade practices, improve traceability, and lessen information asymmetry. Overall, agriculture marketing in India has a wide range of applications, and its effective growth will necessitate a thorough and well-coordinated strategy that tackles the industry's problems and capitalizes on its potential for economic expansion.

Challenges in Agricultural Product Transportation

The primary challenges in transporting agricultural goods in India stem from inadequate infrastructure, insufficient support, and poor logistics connectivity, which hinder farmers' ability to deliver their produce to markets in a timely manner.

- India's transportation system is severely lacking. Only a limited number of villages are connected to mandis via railways or paved roads.
- Produce is often transported using slow and inefficient methods such as bullock carts.
- This mode of transportation is not suitable for covering long distances, forcing farmers to sell their produce in nearby markets even when the prices offered there are far from favorable. This issue is especially prevalent for perishable goods.
- The scarcity of transportation options capable of covering long distances quickly adds to the problem, as these services are both expensive and time-consuming. Moreover, there is a lack of services such as mobile cold storage, which is crucial for maintaining the quality of fresh perishable products that cannot be stored at production sites but need to be transferred quickly.
- Because of the perishable nature of certain fruits, there are limitations on their marketing. These limitations include challenges related to direct marketing, ensuring better returns, dependence on weekly haats, and wholesale markets (mandis).

Initiatives for Enhancing Transportation Systems

Kisan rails are the first multi-commodity trains in the world.

- These chilled freight trains will help in the quick delivery of perishable agricultural products like fruits and vegetables to markets.
- These will make sure that agricultural products are transported across the entire nation.
- The air transport of agricultural commodities proposed by Krishi Udan.
- As a result, value realization (on agricultural items) would greatly increase, especially in the northeast and tribal areas.

The Transport and Marketing Assistance (TMA) programme aims to help with the marketing of agricultural products as well as the international portion of freight.

Kisan Rath mobile application (app) to allow transportation of foodgrains and perishables during the lockdown.

VARIOUS TYPES OF AGRICULTURE MARKETING IN INDIA

Agricultural marketing in India takes various forms, each with its distinct features and advantages. Below are some of the key types of agricultural marketing prevalent in the country:

1. **Traditional Agricultural Marketing:** This refers to the conventional method of selling farm produce in local markets or mandis. In this system, farmers sell their goods directly to intermediaries such as wholesalers and commission agents, who then distribute the products further along the supply chain. Despite being widely practiced, traditional marketing often limits farmers' bargaining power and income potential.
2. **Modern Agricultural Marketing:** Modern marketing incorporates advanced techniques and technological innovations such as e-commerce platforms, supermarkets, and contract farming. This approach provides farmers with better market access and improved price realization. Additionally, modern marketing facilitates the adoption of advanced farming practices, ensuring higher productivity and better-quality produce.
3. **Direct Agricultural Marketing:** This method involves farmers selling their produce directly to consumers through avenues like farmers' markets, farm gate sales, and community-supported agriculture (CSA) initiatives. By eliminating middlemen, direct marketing allows farmers to earn higher profits and build a direct relationship with their customers, fostering trust and transparency in transactions.
4. **Cooperative Agricultural Marketing:** Under this model, farmers collectively establish cooperatives or marketing societies to sell their produce. By pooling their resources and negotiating collectively, farmers can secure better prices, improve market access, and reduce exploitation by intermediaries. Cooperative marketing is particularly beneficial for small and marginal farmers who struggle to compete individually in large markets.
5. **Contract Farming:** In contract farming, agricultural producers enter into agreements with agribusiness firms, food processors, or exporters to cultivate specific crops or rear livestock according to predefined standards. In return, companies provide essential inputs, technical assistance, and assured

market access. This model ensures stable income for farmers while enabling agribusinesses to maintain quality and supply consistency.

6. **Value Chain Agricultural Marketing:** This approach involves integrating all stakeholders in the agricultural supply chain, from production and processing to distribution and retailing. Value chain marketing enhances the efficiency and competitiveness of the agricultural sector by ensuring a seamless flow of products and services. Additionally, it creates new income opportunities for farmers, processors, and distributors, thereby fostering a more resilient agricultural economy.

By adopting diverse agricultural marketing strategies, Indian farmers can enhance their earnings, improve market accessibility, and contribute to the overall development of the agricultural sector.

Obstacles in Transport and Marketing of Agricultural Products

The main issues in the agricultural marketing system are outlined below.

Insufficient Storage Facilities

- Inadequate warehouse facilities are a significant issue in many settlements. Farmers often resort to using trenches, mud containers, "Kutcha" (temporary) storehouses, and other inefficient storage methods, leading to considerable wastage of their produce.
- Approximately 1.5% of the produce goes bad and cannot be consumed, which further floods the local market with excess supply, making it challenging for farmers to secure reasonable prices for their goods.
- However, the situation has somewhat improved with the establishment of the Central Warehousing Corporation and State Warehousing Corporation.

Challenges due to Absence of Standardization and Grading

- Agricultural products are often not graded properly, leading to practices like the "Dara" sales method, where large quantities of goods of different grades are sold in bulk.
- As a result, farmers producing higher-quality goods are uncertain of receiving better prices. This lack of grading discourages the use of superior seeds or the development of improved crop varieties.

Prevalence of Numerous Middlemen

- The agricultural market is heavily dominated by middlemen, which significantly reduces the farmers' share of the profits.

Problems in Unregulated Markets

- India still has many unregulated markets, where unscrupulous middlemen like Arhatiyas and brokers exploit farmers' lack of knowledge and literacy, often defrauding them.

- Farmers are burdened with several hidden and undefined charges, including "arhat" (commission charges), "tulaii" (weight charges), "palledari" (miscellaneous unloading fees), "garda" (impurity charges), and other unlisted fees.

Lack of Market Information

- Farmers often struggle to access accurate market prices from various locations, leading them to accept whatever price is offered by traders.
- To address this, the government regularly broadcasts market prices through radio and television, and print media also provides updates on price changes.

Limited Access to Credit

- Due to financial struggles, Indian farmers often sell their crops immediately after harvest, even when prices are low, to meet their immediate financial needs.

Importance of *Transportation in* Agriculture Produce

Transport is a critical element in the agriculture marketing concerned with delivering agricultural products from farms to markets and to cities worldwide. Transport allows farmers to invest more, increase production and reach the prospective consumers.

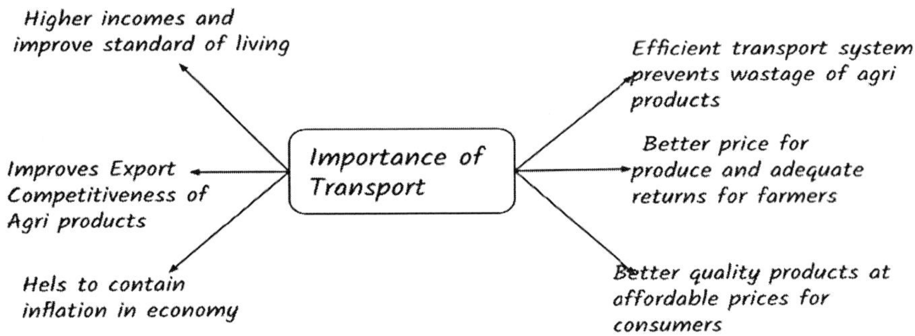

Figure: 7.1 Importance of Transport

Key Challenges in Transport

Government measures to enhance transport infrastructure:

- **PM-Gram Sadak Yojana:** Aimed at improving transport and connectivity in rural India, this initiative focuses on bettering road infrastructure for remote areas.
- **Kisan Rail:** This is India's first-ever multi-commodity train service designed to carry a wide range of agricultural produce, such as vegetables like capsicum, cauliflower, cabbage, chilies, drumsticks, onions, and fruits like grapes, bananas, and pomegranates.
- **Krishi UDAN 2.0:** This initiative incentivizes and supports the air transportation of agricultural products, facilitating quicker and more efficient transport.

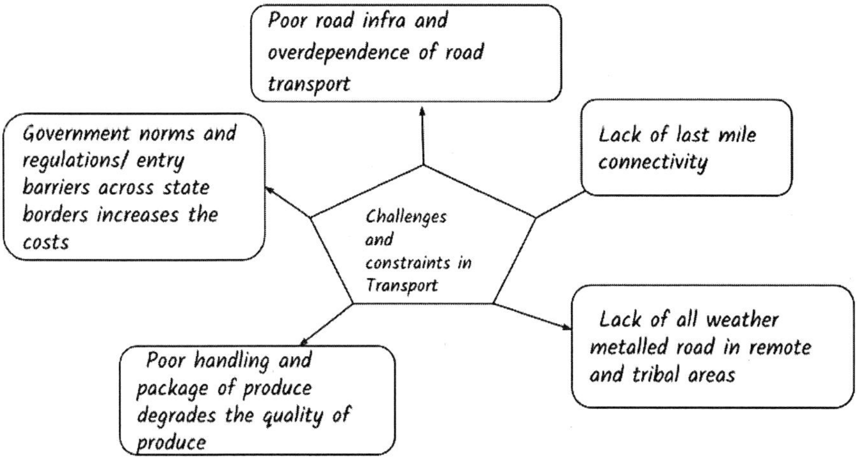

Figure: 7.2 Challenges and constranits in transport

- **Transport and Marketing Assistance (TMA):** This scheme offers financial assistance to exporters of agricultural produce, particularly targeting Europe and North America, by reimbursing a portion of freight charges.
- **Conclusion:** As Henry C. Wallace, a former U.S. Secretary of Agriculture, aptly stated, "Agriculture is our greatest industry, transportation our second greatest. These two industries are interdependent, and the national well-being relies on both."

Transportation in Agriculture Sector

Agricultural transport is considered a vital factor in enhancing agricultural productivity. It plays a crucial role in improving the quality of life, creating markets for agricultural goods, fostering connections between different geographical and economic areas, and opening up new avenues for business. The term "transport" refers to the movement of agricultural goods from one stage of the post-harvest process to the next. For agricultural transport, whether through traditional or mechanized means, transportation is necessary for:

- Moving goods from the harvest fields to threshing or drying areas.
- Transporting them to the farmer's storage facilities or collection-center warehouses.
- Transporting grains from the farm to centralized storage locations.
- Final delivery from storage facilities or industries to wholesalers or retailers for marketing.

Role of Transportation in Agriculture

Transportation allows producers to more effectively collect inputs, move their goods over long distances, and reach international markets with improved infrastructure.

Impact of Transportation on Agriculture

In simple terms, transportation facilitates agricultural activities by enabling farmers to invest more and increase their output. Without an efficient transport system, vast amounts of carefully cultivated produce would be wasted. Additionally, transportation ensures that the produce reaches consumers at a reasonable cost and within an appropriate timeframe.

Figure: 7.3 Transportation in Agriculture

Traditional Transport

In many parts of the world, farmers and peasants live in remote areas far from transport routes where their grain can be moved to collection, storage, or marketing facilities. As a result, agricultural products are often transported in small quantities over poorly maintained roads or pathways. This leads to long delays in delivery and high transportation costs per unit of product, significantly reducing the farmers' income. Such conditions discourage farmers from increasing their production.

To increase output, it is essential not only to build adequate storage facilities but also to improve and realign local road networks to accommodate the transportation of products. In areas with poor road infrastructure and traditional agricultural transport methods, products are typically moved by people, donkeys, camels, and occasionally horses, with oxen being the most common draft animals.

In other cases, products are transported to collection points, storage, and marketing locations using pickup trucks, buses, and taxis. In many instances, farmers are required to travel distances of 30 or 40 kilometers to collect, store, or market their produce in locations where transportation is still largely carried out by people or animals.

A rental system often exists for animals or small trucks, with the rental charges varying depending on the season, road conditions, and distance to be covered. Programs aimed at promoting the use of oxcarts and basic wheelbarrows have a higher chance of success when the equipment can be locally manufactured by artisans. However, such initiatives often fail when imported parts like axles and wheels cannot be repaired locally.

TECHNIQUES FOR MARKETING SUSTAINABLE AGRICULTURE

1. Market intelligence and information systems: Farmers can obtain up-to-date data on consumer preferences, prices, and market trends by putting in place strong market intelligence systems. This increases their competitiveness in the market by enabling them to make well-informed decisions.
2. Direct marketing and farmer producer organizations (FPOS): Farmers can avoid using too many middlemen by supporting direct marketing campaigns and the establishment of farmer producer organizations. FPOS gives small-scale farmers better access to markets, direct sales, and collective bargaining.
3. Value addition and branding: Agricultural products' market appeal is increased by investing in value addition through processing and branding. This raises farmers' income by enabling them to demand higher prices for goods with added value.
4. Infrastructure investment: Governments and interested parties must make investments to upgrade the infrastructure for processing, transportation, and storage. Modern processing facilities, effective transportation systems, and cold storage facilities all help to guarantee product quality and minimize post-harvest losses.
5. Contract farming and agri-logistics: Supply chains can become more organized and effective by encouraging contract farming agreements and improving agri-logistics. While better logistics guarantee prompt and secure product transportation, contract farming offers demand stability.
6. Technology integration: Operations can be streamlined by incorporating technology into agricultural marketing procedures. Farmers can reach a larger market and establish direct connections with consumers through e-commerce solutions, mobile apps, and online platforms.
7. Government support and policies: The resilience of the agriculture marketing industry can be greatly increased by supportive government policies that address market challenges and offer financial aid, subsidies, and incentives.

A vital part of the agri-food system, agriculture marketing links farmers with consumers and influences the economy. Addressing the issues that farmers and other stakeholders face is crucial to ensuring this sector grows sustainably. All parties involved in the agricultural value chain stand to gain from a more effective, transparent, and resilient agriculture marketing system that is made possible by putting creative ideas into practice, making infrastructural investments, and utilizing technology.

In order to give farmers stability and financial support so they can invest in modernization and expansion projects, agricultural loan programs must be implemented effectively. The development of efficient marketing techniques becomes essential for the expansion and sustainability of the agricultural industry as the demand for agricultural products rises globally.

Criteria for Agriculture Marketing

From small farmers to major agribusinesses, a variety of stakeholders are involved in agriculture marketing in India. Certain eligibility requirements must be fulfilled in order to take part in formal agriculture marketing systems. These standards guarantee that the procedure is open, effective, and advantageous to all stakeholders. The main qualifying criteria are listed below:

1. Farmers' and sellers' registration: Farmers or producers must register with the appropriate local authorities or marketing boards in order to participate in formal agriculture markets, such as mandis or e-NAM. This guarantees that these platforms are only accessible by authorized sellers.
2. Licensing for buyers and traders: The state Agricultural Produce Market Committee (APMC) issues valid licenses to buyers, wholesalers, and traders who operate in agricultural markets. This is to guarantee that purchasers are reliable, well-off, and able to pay fairly for the produce.
3. Adherence to quality standards: Farmers are required to meet particular quality standards established by regulatory organizations in order to market their produce. Depending on the type of produce, these requirements may change, but they usually include post-harvest protocols, appropriate grading, and hygiene maintenance.
4. Participation in mandis or e-NAM: Farmers must sell their produce online at e-NAM or at authorized mandis, or agriculture markets. To participate in online sales, farmers in the e-NAM case should have access to digital tools, banking, and transportation facilities.
5. Traceability and record keeping: Farmers are urged to keep accurate records of their agricultural activities, including the kinds of inputs they use, the methods they employ for harvesting, and other pertinent details. This guarantees produce traceability, which is frequently necessary for export markets and quality certifications.
6. Compliance with market rules and regulations: Farmers and buyers must abide by market rules pertaining to pricing, auction procedures, and product weighing in order to be eligible for formal marketing systems. Laws against hoarding and collusion must also be followed.
7. Infrastructurerequirements for sellers: Farmers should have the necessary infrastructure, such as proper packaging, transportation, and storage, to participate effectively in marketing their produce. This is particularly important for perishable goods, as they may need to be stored cold or transported in a refrigerated environment.

8. Cooperative memberships: To access opportunities for collective marketing, farmers may occasionally need to be a member of cooperative societies or farmer producer organizations (FPOs). By pooling produce and negotiating as a group, these organizations frequently assist in obtaining better prices.
9. Financial and legal responsibilities: Farmers and traders involved in agriculture marketing have financial and legal responsibilities, such as paying taxes and market fees and adhering to applicable government programs and subsidies.
10. Access to digital and physical marketplaces: To get the most out of contemporary agriculture marketing, farmers must have access to both digital and physical marketing platforms. This frequently entails possessing mobile phones, bank accounts, and knowledge of online resources such as e-NAM or other government-run programs.

Reforms to agriculture marketing, like the creation of e-NAM and modifications to the APMC Acts, have attempted to increase the processes' accessibility and inclusivity for a larger group of Indian farmers.

India's Agriculture Marketing Advantages

In India, agricultural marketing is crucial to the sector's productivity and financial success. It affects farmers' incomes and the economy as a whole, and it has a big influence on how agricultural products are distributed. Here are a few main advantages:

1. Increased income for farmers: Agriculture marketing gives farmers access to a wider range of markets, including both domestic and foreign consumers. This directly raises their income levels by increasing their chances of receiving better prices for their produce.
2. Decrease in post-harvest losses: Timely transportation, storage, and sale of agricultural products are made possible by effective marketing channels, which lowers post-harvest losses. Effective marketing strategies reduce waste and enhance the caliber of products that are delivered to customers.
3. Price discovery and equitable returns: Through transparent auctions, agricultural marketing platforms like e-NAM (National Agriculture Market) and other mandis (markets) facilitate improved price discovery. This makes it less likely that middlemen will take advantage of farmers and guarantees that they receive a just price for their produce.
4. Access to improved inputs and technology: Farmers who participate in organized marketing systems are more likely to be exposed to new equipment, fertilizers, pesticides, and farming technologies that raise produce quality and productivity.
5. Better marketing networks allow farmers to build their credibility, which in turn makes credit and financial products like loans, insurance, and subsidies more accessible. This enables them to grow their farming operations and control risks.
6. Crop and farming practice diversification: By connecting farmers to markets where there is a greater need for a range of crops, agriculture marketing

systems promote crop diversification. This encourages sustainable farming methods and lessens dependency on conventional crops.

7. Promotion of agro-based industries: By offering a consistent supply of raw materials, robust agriculture marketing frameworks support the expansion of agro-based industries. This helps the local economy grow and create jobs in rural areas.
8. Higher exports: Indian agricultural products are exported more frequently thanks to agriculture marketing, especially through foreign channels. This increases foreign exchange reserves and improves the nation's trade balance.
9. Informational empowerment of farmers: Marketing platforms provide farmers with essential data on demand, prices, quality standards, and market trends, empowering them to make well-informed choices about their production and sales plans.

Marketing Functions

A little earlier it was said that a marketing system has two distinct dimensions. One of those dimensions is the institutions, organisations and enterprises which participate in a market and the second is the functions that those participants perform.

A. Exchange Functions	1. Buying
	2. Selling
B. Physical Functions	3. Storage
	4. Transportation
	5. Processing
C. Facilitating Functions	6. Standardisation
	7. Financing
	8. Risk Bearing
	9. Market Intelligence

Each of these functions add value to the product and they require inputs, so they incur costs. As long as the value added to the product is positive, most firms or entrepreneurs will find it profitable to compete to supply the service.

Exchange Functions

- **Buying:** The marketing concept emphasizes the importance of understanding and meeting customer needs. A producer adopts a market orientation when their production is specifically planned to fulfill defined market demands or opportunities. For instance, a contract farmer looking to meet the requirements of a food processor producing sorghum-based malted drinks will only purchase improved sorghum seed. They will avoid any inputs that could negatively impact the sorghum's storage or processing properties and will continually seek higher-quality inputs to add more value to the product in the eyes of the customer. When making purchasing decisions, the farmer's primary concern will be how these choices affect the marketability of their product. The buyer's motivation is typically driven by the desire to maintain or increase profits, rather than necessarily to provide the best quality. Enhancing quality often raises costs, and in some cases, consumers may not be willing to pay a premium for quality improvements beyond a certain level. In such scenarios, a grower focused solely on producing a "better product" might not be market-oriented, as they are disregarding the actual needs of the consumer. The most successful agribusiness is one that maximizes the difference between the prices it can charge and the costs it incurs.

- **Selling:** This is one of the functions most closely associated with marketing, and for many, the terms marketing and selling are often viewed as the same. Kotler suggests that: "Most firms practice the selling concept when they have overcapacity. Their immediate aim is to sell what they can make rather than to make what they can sell." There is no doubt that high-pressure selling is a practice where the interests of the consumer are not the top priority of the seller. However, this does not constitute marketing. The marketing philosophy emerges when firms realize that their long-term objectives can only be achieved through consistently providing customer satisfaction. While selling can create a consumer, marketing is about creating a lasting customer. The key difference is that marketing focuses on establishing and nurturing long-term customer relationships. Selling is a component of marketing, much like promotion, advertising, and merchandising, all of which are part of the broader marketing mix. These elements are focused on persuasion and collectively fall under the category of marketing communications, one of the four key components of the marketing mix.

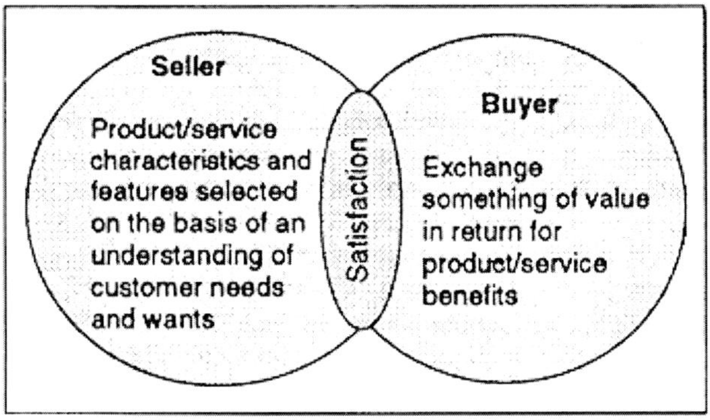

Figure: 7.4 The Exchange Function

Physical Functions

Storage: One of the inherent characteristics of agricultural production is its seasonality, whereas demand for agricultural products tends to be continuous throughout the year. This mismatch between production and demand creates the need for storage to ensure a smooth and continuous flow of goods into the market.

Unlike manufacturers, who can adjust production timing to align with demand, agricultural producers have limited flexibility. While manufacturers too face production constraints, they have more options than agricultural producers. For example, manufacturers can work overtime, subcontract work, or even adjust their production capacity over time to align with market demand. In many developing countries (LDCs), there is often an oversupply of agricultural products immediately after harvest, leading to reduced prices and significant waste. Conversely, before the next harvest, supply can become scarce, and traders and consumers must pay premium prices for the limited available produce. The storage function plays a key role in balancing supply and demand. Both farmers and consumers benefit from a marketing system that ensures products are available when needed. Those involved in storage—whether farmers, merchants, co-operatives, marketing boards, or retailers—provide a valuable service. This service, however, comes at a cost, and there are risks such as spoilage and fluctuations in demand or prices. Thus, those providing storage services are entitled to a reward, typically in the form of profit.

Transportation: The transportation function's primary role is to ensure that products are delivered to the required locations without significantly increasing the overall cost of the goods. For effective transportation, it is necessary to consider alternative routes and modes of transport to ensure timely delivery, maintain product quality, and minimize shipping expenses. Efficient transportation management is crucial for successful marketing. Whether operating a single vehicle or a fleet, transportation must be meticulously managed, which includes monitoring costs, managing operations across different road types, and overseeing fuel consumption, lubrication, and vehicle maintenance. Skillful management of vehicle operations can significantly enhance marketing efficiency, particularly through optimal routing, scheduling, and loading/unloading processes. Additionally, it involves maximizing available shift hours, maintaining the vehicle fleet at an optimal size, and carefully managing delivery and collection schedules, as well as making informed decisions regarding vehicle replacement and depreciation. Transport managers must also assess the pros and cons of owning, leasing, or hiring transport vehicles.

Processing: Most agricultural products are not immediately suitable for direct consumption once harvested. They typically require some form of processing before they can be used. However, it is precisely because of this form-changing aspect that processing should be considered a marketing function. Changing the form of a product, such as converting green coffee beans into roasted beans, cassava into gari, or livestock feed into processed feed, adds value to the product. This transformation makes the product more useful and desirable to buyers. Decisions regarding how and what form changes to apply to a product are marketing decisions. Similarly, a fresh fruit producer with pulping or canning facilities might face the issue of reducing product flexibility if buyers prefer to use the fruit in various forms. In such cases, processing may actually decrease the product's utility and value. Processing is not the only means of adding value to a product. Storing products until they are needed adds utility and value, as does transporting goods to locations convenient for consumers. In essence, any action that enhances the utility of a product or service for potential buyers contributes to its value.

Facilitating Functions

The facilitating functions include product standardisation, financing, risk bearing and market intelligence. Facilitating functions are those activities which enable the exchange process to take place. Marketing, in simple terms, is the act of supplying products to someone in exchange for something perceived to be of equal or greater value, (usually, but not always, a given sum of money). Facilitating functions are not a direct part of either the exchange of title or the physical movement of produce.

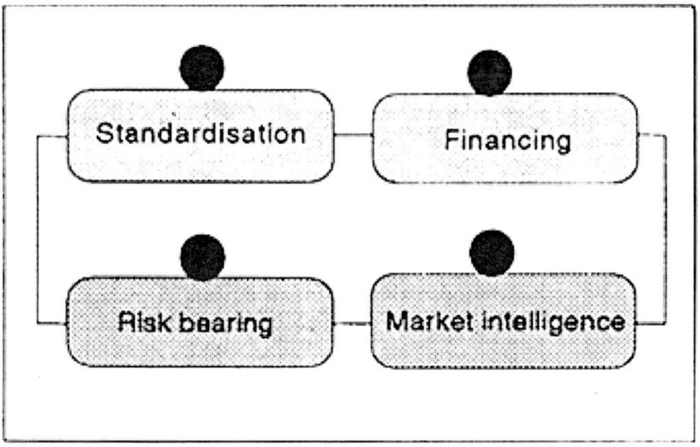

Figure: 7.5 The Facilitating Punctions

Standardisation: Standardization refers to the process of establishing and maintaining consistent measurements for the quality and/or quantity of produce. This function simplifies the buying and selling process, reducing marketing costs by allowing buyers to specify exactly what they require, while suppliers can clearly communicate what they are capable of offering in terms of both quantity and quality. Without standard weights and measures, trade becomes either more expensive or even impossible. In Nepal, for example, the wide variety of weights and measures used for grain led to a situation where it was easier for certain districts to trade with neighboring states in India than with other districts within Nepal itself. The key benefits of having uniform standards include:

- Price quotations becoming more meaningful and comparable.
- The ability to sell commodities by sample or description.
- Small quantities of products from many small producers can be aggregated into larger, economically viable loads, provided the supplies are of similar quality or grade.
- When faced with a range of graded products, buyers can select the quality of product they are willing and able to purchase.

Quality differences in agricultural products arise for several reasons. These can be due to varying production methods or the quality of inputs used. Technological advancements can also lead to differences in product quality. Additionally, a buyer's evaluation of a product's quality is often influenced by personal preferences. For example, in some markets, a smaller banana may be perceived as superior to a larger one; white sugar is considered to be of higher quality than yellow sugar; long-stemmed carnations are seen as higher quality than short-stemmed ones; and white maize is thought to be easier to digest than yellow maize. Whether the criteria for assessing quality are objective or subjective does not change their impact in the marketplace. What matters in marketing is understanding how the buyer perceives and evaluates 'quality'.

Financing: In nearly every production system, there are unavoidable delays between investing in necessary raw materials (such as machinery, seeds, fertilizers, packaging, flavorings, stock, etc.) and receiving payment from the sale of the finished product. During these periods of delay, some individual or organization must finance the initial investment. The issue of where the funding for these investments will come from, at every stage between production and consumption, is a critical consideration that marketing must address. Take, for example, a food manufacturer who wants to introduce a range of chilled products in a developing country where few retail outlets have the necessary refrigeration facilities. This is a marketing challenge. It could be solved by the food manufacturer purchasing refrigerators and leasing them to retailers, or setting up hire-purchase arrangements with the retailers.

In developing countries, a frequent marketing problem is the low income levels, which lead to insufficient effective demand for many products. The marketing challenge here is to find ways to channel the available income into meaningful demand. In the case of agricultural equipment marketing, this may involve offering hire-purchase schemes, where the buyer makes regular installment payments. During this period, the buyer is considered to be renting the equipment. If the payments stop, the equipment can be reclaimed, as ownership remains with the seller until the full payment is made, at which point the farmer officially owns the equipment. Alternatively, the seller may offer leasing arrangements instead of purchase schemes, where the farmer makes regular payments but never owns the equipment. For food items targeted at low-income markets, the seller might reduce the unit price of the product by offering smaller pack sizes. Another strategy could involve using cheaper ingredients or packaging. For example, instant coffee can be made more affordable by substituting part of the coffee with chicory, or the price of meat products can be lowered by increasing the percentage of cereals in the product and using less expensive parts of the animal such as entrails, offal, feet, and head.

Marketing is also involved in securing financing for the enterprise itself. Creative solutions can be developed in this area as well. When internal financing is not sufficient, businesses in developing countries have several alternative sources of funding, including:

- Development banks
- Commercial banks
- Stock market shares
- Credit cooperatives or credit unions

If these financial options are unsuitable or unavailable, a strategic alliance, such as a joint venture, may be the solution. Joint ventures are partnerships formed to better exploit market opportunities in ways that neither party could achieve alone. A business in a developing country might form a joint venture with a local or foreign partner. The joint venture agreement typically outlines each party's contributions of resources, their share of management control, and their respective share of profit and risk.

Links Between Agriculture and the Food Industry

The connection between agriculture and food is constantly evolving. In early societies, the farmer and the consumer were often the same person or closely related neighbors who exchanged goods and services directly, as shown in figure 1.1. However, as societies progressed, additional links were introduced. These include commodity traders, processors, manufacturers who transform raw products into food, and retailers, among others, who now serve as intermediaries between the producer and the consumer. More recently, scientists have become an integral part of this chain. Breeders, plant biologists, nutritionists, and chemists have made invaluable contributions to agricultural production and food manufacturing over the past five decades. It seems that we have moved beyond the era dominated by machines in agriculture and chemicals, and have entered the age of biotechnology. Biotechnology holds great promise for developing countries, as it tends to be less capital-intensive and more reliant on research and expertise. This means that the benefits of biotechnology could reach poorer countries faster, as they often lack the capital required for more traditional advancements. As a result, the impact of biotechnology could be more rapid, widespread, and profound.

As disposable incomes increase, the food industry will increase the quality and diversity of the products it produces. Food manufacturers will have particular expectations of agriculture as a supplier of their raw materials, including:

- **Quality:** To build a successful business, food manufacturers aim to create a preference for their products by differentiating them in ways that are meaningful to consumers. To help consumers recognize these unique products, manufacturers brand them and then work to foster loyalty to those brands. However, brand loyalty is typically achieved only by consistently delivering high-quality products. As disposable incomes rise, consumer demands become more sophisticated, and the quality of raw materials becomes increasingly important. When agriculture serves a food industry that must meet these more complex demands, it will face a growing emphasis on quality. Likewise, agriculture will benefit from higher returns for innovations that improve product quality.

- **Cost:** After quality, cost is a major consideration. With better access to global raw material sources, the food industry can find the lowest cost supplier that still meets the required quality. For food manufacturers, the country of production or marketing is no longer the sole source of agricultural produce. Enhanced transportation and communication have made the entire world a potential source of supply. This shift represents a significant change in agriculture's competitive environment, which has largely been centered around local markets.
- **Non-seasonality:** Historically, agricultural products followed seasonal production and supply cycles. However, modern technology and improved farming practices allow food manufacturers to operate without being tied to seasonal production. The capital-intensive nature of the food industry means it cannot afford the costs of underutilizing its production capacity. As a result, farmers must work to reduce seasonality or adapt to competitive cycles in production and supply.
- **Reliability:** A manufacturer who invests heavily in building a brand will want reliable supplies of agricultural products in terms of quality, timing, and cost. Agricultural producers will increasingly be judged on their ability to meet these reliability standards.
- **Processing:** The ease of processing will become a growing expectation for the food industry. As in all industries, reducing the costs of capital equipment, labor, and inventory is essential. For instance, farmers who can provide goods based on the "just-in-time" principle will help reduce manufacturers' working capital and space needs. Farmers who participate in secondary processing or perform tasks such as post-harvest treatments or transportation will offer added value. Additionally, crops designed to simplify processing, such as seedless fruits or low-cholesterol meats, will benefit the food industry. In summary, competitive advantage will lie with those who can add the most value and differentiate their offerings from other suppliers.
- **Product Differentiation:** In competitive brand marketing, the food industry must continuously innovate to create new products that surpass existing ones, either from their own brand or competitors. While innovation will remain vital at the processing stage, manufacturers will increasingly look for innovative changes in the agricultural products themselves. These innovations may involve new tastes, better textures, or more appealing shapes.
- **Health Aspects:** In more advanced food markets, healthy eating is becoming a higher priority among consumers. Therefore, farmers must consider the health implications of the crops they grow. There are two key aspects of health to consider. First, consumers may prioritize the food itself, seeking low-fat, low/no sugar, or low/no salt options. Health concerns are not limited to more affluent markets or segments, as even the poor, while receiving enough food to avoid starvation, may still suffer from malnutrition. Farmers must thus be mindful of the nutritional value of their crops. Second, consumers may be concerned about food production methods, such as the use of chemicals

like herbicides and pesticides. This may require changes to farming practices, impacting production costs. Both consumers and the food industry will expect farmers to grow products without potentially harmful chemicals, yet without increasing costs, posing a significant challenge for agriculture.

AGRICULTURAL AND FOOD MARKETING ENTERPRISES IN INDIA

The agricultural and food marketing industries in India, one of the world's biggest producers, are extensive and constantly changing. In order to guarantee that farm produce reaches consumers effectively, increase farmers' income, and advance food security, agricultural and food marketing businesses are essential.

These businesses work in a variety of sectors, such as manufacturing, processing, distribution, retail, and exporting. Technology breakthroughs, digital platforms, and policy changes have all contributed to the notable expansion and change of India's agriculture and food marketing industry.

Types of Agricultural and Food Marketing Enterprises in India

1. **Government-Owned Enterprises:** Government organizations such as the Food Corporation of India (FCI), National Agricultural Cooperative Marketing Federation of India (NAFED), and the Agricultural Produce Market Committees (APMCs) play a crucial role in food procurement, storage, and distribution. These enterprises regulate market operations, stabilize prices, and support farmers by offering minimum support prices (MSP) for key agricultural commodities.
2. **Cooperative Marketing Enterprises:** Agricultural cooperatives, such as Amul (for dairy products) and Indian Farmers Fertiliser Cooperative Limited (IFFCO), help farmers collectively market their produce. These enterprises enable small and marginal farmers to gain better market access, improve bargaining power, and ensure fair pricing for their products. Cooperative marketing enterprises have been instrumental in transforming India's dairy and fertilizer industries.
3. **Private Agribusiness Companies:** Private sector enterprises such as ITC, Reliance Fresh, Mahindra Agribusiness, and Adani Agri Logistics have played a significant role in modernizing agricultural marketing. These companies provide farmers with better access to markets, quality inputs, and storage facilities. Some of them operate farm-to-fork models, ensuring fair pricing and reducing post-harvest losses.
4. **Food Processing and Packaging Enterprises:** Companies involved in food processing, such as Nestlé India, Britannia Industries, and Haldiram's, source raw agricultural products from farmers and convert them into value-added products. These enterprises enhance food safety, extend shelf life, and increase the economic value of agricultural produce. The growth of food processing industries has also contributed to increased employment opportunities in rural areas.

5. **E-Commerce and Agri-Tech Startups:** The rise of digital technology has led to the emergence of agri-tech startups and e-commerce platforms such as Ninjacart, DeHaat, BigBasket, and WayCool. These enterprises connect farmers directly with consumers, retailers, and institutional buyers through digital marketplaces, ensuring better price realization and reducing dependence on traditional supply chains. They also offer AI-driven analytics, precision farming solutions, and access to farm inputs.
6. **Export-Oriented Enterprises:** India is a major exporter of agricultural and food products, with companies such as LT Foods (Daawat), KRBL (India Gate Basmati), and various spice exporters playing a crucial role in global trade. Organizations like the Agricultural and Processed Food Products Export Development Authority (APEDA) facilitate the export of Indian agricultural commodities by ensuring quality compliance and international market linkages.

Challenges Faced by Agricultural and Food Marketing Enterprises

Despite their growth, agricultural and food marketing enterprises in India face several challenges:

- **Supply Chain Inefficiencies:** Lack of proper storage, transportation, and cold chain infrastructure leads to high post-harvest losses.
- **Market Fluctuations:** Price volatility and unpredictable market conditions affect both farmers and businesses.
- **Regulatory Barriers:** Complex regulations, licensing requirements, and policy uncertainties hinder the smooth operation of agribusinesses.
- **Smallholder Farmer Participation:** Many small and marginal farmers struggle to integrate into organized marketing channels due to a lack of awareness and access to resources.

Future Prospects and Opportunities

With the Indian government promoting agricultural reforms, such as the establishment of farmer-producer organizations (FPOs), digital marketplaces, and financial incentives for food processing industries, the agricultural and food marketing sector is expected to grow further. Increased investments in agri-tech, sustainable farming, and global trade will drive innovation and efficiency in the sector, ultimately benefiting farmers and consumers alike.

Agricultural and food marketing enterprises in India have a pivotal role in transforming the agricultural economy, ensuring food security, and boosting rural incomes. With continued advancements, these enterprises will be key contributors to India's goal of becoming a global leader in agricultural trade and food production.

Agricultural Marketing Boards in India

Agricultural marketing boards in India play a crucial role in regulating, promoting, and facilitating the sale of agricultural produce across the country.

These boards ensure fair pricing, market stability, and efficient distribution channels for farmers, traders, and consumers. By implementing policies, infrastructure development, and quality control measures, these boards contribute to the overall growth of the agricultural sector.

India's agricultural marketing system involves a combination of government-regulated boards, cooperative organizations, and private sector participation. These boards are responsible for managing agricultural markets (mandis), overseeing procurement operations, and facilitating export opportunities for Indian farmers.

Major Agricultural Marketing Boards in India

1. **Agricultural and Processed Food Products Export Development Authority (APEDA):** APEDA is a government agency responsible for the promotion of agricultural exports, particularly processed food products. Established in 1986 under the Ministry of Commerce and Industry, APEDA provides financial assistance, training, and advisory services to exporters. It also ensures compliance with international quality standards to enhance India's competitiveness in global markets. Key focus areas include horticultural products, dairy, meat, and organic food items.

2. **Food Corporation of India (FCI):** Established in 1965, the Food Corporation of India plays a vital role in India's food security by managing the procurement, storage, and distribution of essential grains like wheat and rice. It procures food grains from farmers at Minimum Support Prices (MSP) and distributes them through the Public Distribution System (PDS). FCI helps stabilize food prices, prevent food shortages, and ensure that farmers receive fair compensation for their produce.

3. **National Agricultural Cooperative Marketing Federation of India (NAFED):** NAFED is a cooperative marketing organization that works to safeguard farmers' interests by ensuring fair prices for their agricultural produce. Established in 1958, it operates a network of cooperative societies that procure, store, and sell commodities such as oilseeds, pulses, fruits, and vegetables. NAFED also supports government initiatives like buffer stock management and price stabilization measures.

4. **State Agricultural Marketing Boards (SAMBs):** Each Indian state has its own Agricultural Marketing Board responsible for regulating local agricultural markets and mandis. These boards oversee the functioning of Agricultural Produce Market Committees (APMCs), ensuring transparency, fair pricing, and infrastructural development. They facilitate market linkages for farmers and promote modern marketing techniques like e-NAM (electronic National Agriculture Market). Examples include:

- Punjab State Agricultural Marketing Board
- Maharashtra State Agricultural Marketing Board
- Gujarat State Agricultural Marketing Board

5. **Spices Board of India:** The Spices Board of India, under the Ministry of Commerce and Industry, is responsible for the development and promotion of

Indian spices in both domestic and international markets. It supports spice farmers by providing training, research, quality control, and export assistance. India is one of the largest exporters of spices, and the board plays a crucial role in maintaining quality standards and boosting global demand.

6. Cotton Corporation of India (CCI): The Cotton Corporation of India is a government agency that helps cotton farmers by procuring cotton at MSP and ensuring fair trade practices. It stabilizes cotton prices in the domestic market and assists in export activities. CCI also promotes research and development in cotton cultivation and processing.

7. Tea Board of India: The Tea Board of India, established in 1953, oversees the promotion, research, and export of Indian tea. It provides financial and technical assistance to tea growers, especially small-scale farmers. India is one of the largest tea producers in the world, and the Tea Board ensures the industry's sustainable growth through regulatory measures and international marketing efforts.

8. Coffee Board of India: The Coffee Board of India promotes coffee production, research, and exports. It provides financial support to coffee growers, especially in Karnataka, Kerala, and Tamil Nadu. The board helps small farmers access global markets and encourages sustainable coffee farming practices.

Marketing Boards in Developing Countries

Marketing boards are typically government agencies or statutory organizations that intervene in the marketing process to ensure efficient and organized marketing. Less commonly, they are voluntary organizations formed by farmers or producers. These boards are usually created through government policy rather than by a consensus among commercial parties. This is particularly true for marketing boards in tropical regions, where the primary goal is to enhance the income of smallholders, growers, and livestock farmers. Marketing boards typically do not offer marketing services to large estates or plantations. Before the implementation of structural adjustment and market liberalization, nearly all marketing boards functioned as 'price stabilizing boards'.

Another distinguishing feature of marketing boards is their focus on durable products. They are typically authorized to manage 'controlled' or 'scheduled crops'. In many countries, fewer than five crops are controlled, which often include traditional crops like millet, sorghum, rice, maize, groundnuts, and palm oil, as well as 'colonial' crops like cocoa, cotton, coffee, tea, tobacco, and rubber. Some governments have established boards that manage multiple crops. In certain cases, the marketing board handles all the marketing functions independently, while in others, it partners with private enterprises by renting storage facilities or appointing local buying agents.

The effectiveness of a marketing board is generally assessed based on below key factors:

- Its contribution to ensuring orderly and efficient marketing.
- The reduction in the ability of intermediaries to manipulate profit margins to the detriment of producers and consumers.
- The creation of monopoly power that benefits producers.

In many instances, the establishment of a marketing board was a response to situations where middlemen or foreign buyers were perceived to exert monopsonistic power over producers. As a result, the role of marketing boards is often framed as organizing producers into monopolistic entities with real countervailing power, aiming to reduce inefficiencies caused by unnecessary competition and duplication of efforts among intermediaries.

Key Roles Played by Marketing Boards

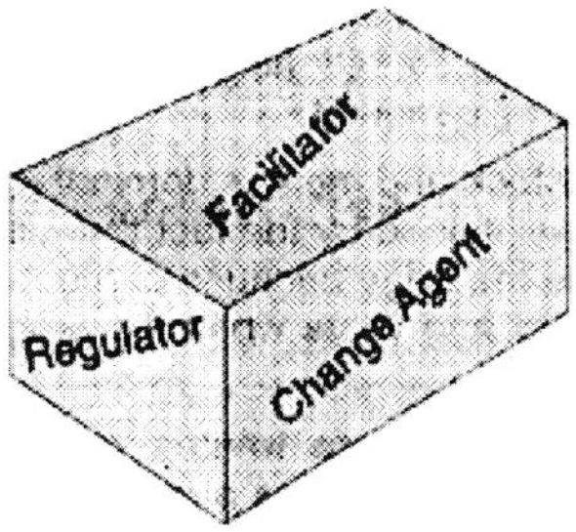

Figure: 7.6 Different Roles of Marketing Board

In theory at least, the marketing board contributes to orderly marketing by acting as an agent for improving marketing practices, as a market regulator and as a provider of facilitating services. For instance:

Change agent	Marketing boards can establish marketing practices and procedures for raw and/or processed products.
Regulatory role	Marketing boards may act as "watch-dogs" over agreed marketing practices and procedures e.g. credit arrangements, weights and measures, quality control etc.
Facilitator	Marketing boards may provide all or some of the facilitating services e.g. credit, market intelligence and risk management. The last of these usually takes the form of the guaranteeing of prices. In the case of tree crops prices are announced in advance of harvest. Prices for annual crops are normally made known before planting or sowing.

The role of marketing boards in promoting more efficient marketing is often presented by policymakers as a means to modify the existing market structure, aiming to transform what is seen as an imperfect market into one that benefits producers. However, when attempting to do this, it's essential to consider the impact on both consumers and other stakeholders in the marketing system. Unfortunately, this consideration is not always made, and it raises the question of whether a market structure that primarily benefits one group of participants can truly be considered beneficial for others.

The argument for granting producers real countervailing power is particularly compelling in scenarios where the marketing system is dominated by many relatively powerless producers and a small number of powerful intermediaries. In such cases, the intermediaries become the price-setters, while producers and consumers are left as price-takers.

A specific way a marketing board might influence the market structure is by streamlining the system to reduce inefficiencies attributed to unnecessary competition and the duplication of efforts between intermediaries. For instance, there could be overlapping transport, storage, and processing facilities that prevent capacity utilization from reaching economic levels without incurring excessively high costs. To address these inefficiencies, marketing boards may introduce measures like licensing systems to help rationalize the market.

Buying Operations of Marketing Boards

Marketing boards typically purchase produce at fixed prices, with the government setting these prices for scheduled crops each season or year. For tree crops, the price is announced before harvest, and for annual crops, it is set before planting or sowing. The price remains stable for a specific period, usually around six months, offering producers a sense of security.

Purchases occur at official buying points, which are equipped with either storage facilities or transportation systems to move the produce before its quality deteriorates. Farmers naturally prefer that these buying points be conveniently located. However, maintaining an extensive network of such points significantly increases the marketing board's operational costs, which can lead to conflicting interests between the board and the farmers. One compromise is for the marketing board to use mobile buying teams in addition to permanent buying points.

In some regions, the buying points are staffed by marketing board employees. However, the high operational and transportation costs may lead governments to seek alternative solutions, such as transferring the responsibility of buying points to local cooperatives or appointing licensed buying agents (LBAs).

Challenges Faced by Agricultural Marketing Boards in India

Agricultural Marketing Boards' Obstacles Agricultural marketing boards in India face a number of difficulties despite their significance:

- **Market Inefficiencies:** Farmers frequently receive smaller returns when there are several middlemen.
- **Infrastructure Gaps:** Post-harvest losses are caused by inadequate cold chain, storage, and transportation facilities.
- **Regulatory Issues:** In some states, there is opposition to the adoption of marketing reforms like APMC liberalization.
- **Digital Integration:** For clear and effective trading, many traditional mandis and marketing boards have not yet fully embraced digital platforms.

Future Prospects and Reforms

To enhance agricultural marketing, the Indian government has introduced several reforms, including:

- **e-NAM (National Agriculture Market):** A digital trading platform that connects mandis across India to create a unified market.
- **Promotion of Farmer Producer Organizations (FPOs):** Strengthening collective bargaining power for farmers.
- **Improvement of Logistics and Cold Storage:** Investment in better storage and transportation infrastructure to reduce wastage.

In India, agricultural marketing boards play a crucial role in aiding farmers, guaranteeing food security, and fostering exports. A more effective and farmer-friendly marketing system will result from strengthening these boards through infrastructure development, digital adoption, and policy reforms.

Selling Operations of Marketing Boards

Certain marketing boards, such as those dealing with grain, focus solely on domestic consumer markets. These boards typically handle essential crops like maize, millet, and rice. In contrast, some boards focus entirely on export markets, catering to industrial buyers. The nature of these two types of markets is distinctly different, and as a result, the operations of the boards serving each market differ accordingly. To distinguish between these two types, they are sometimes referred to as Food Marketing Boards (FMBs) and Export Marketing Boards (EMBs). A significant difference between them lies in the role of governments. Governments typically have no influence over the demand in export markets, while they can and do control demand within the domestic market. Furthermore since governments must consider the interests of domestic consumers of staple crops, they sometimes direct FMBs to adjust their marketing strategies to align with social or political goals, rather than purely commercial objectives. In contrast, the government of the exporting country does not have the same responsibility to manage the interests of consumers in foreign markets.

- **Selling operations of EMBs:** Some export markets are regulated by commodity agreements such as the Sugar, Cocoa, and International Coffee Agreements. However, in most cases, they must operate in free or open markets where intense competition is prevalent. EMBs generally prefer early sales, aiming to minimize the time between buying and exporting, which is often referred to as a "rapid evacuation" policy. This strategy minimizes storage and capital investment needs since the recipient of the produce takes on the responsibility for holding and financing the stocks. Most EMBs engage in "forward selling," meaning they sign sales contracts well ahead of delivery. This can involve selling crops even before they are harvested or, in some cases, before they are even planted.

  A common practice among EMBs is "export parity pricing," where the producer price is calculated as the export price minus marketing costs. There is little incentive to reduce these marketing costs within such a system, as EMBs are primarily concerned with the instability of prices in open world markets. Since they cannot control these prices, EMBs typically adopt a defensive stance by advocating for lower producer prices to avoid trading deficits when world prices decline.

- **Selling operations of FMBs:** In many developing countries, the selling price of Food Marketing Boards (FMBs) is determined by the government. Governments, often motivated by the desire to protect consumers, tend to set low prices. As a result, the gross trading margin for FMBs is usually quite small, and this margin frequently causes tension between FMBs and the government. Governments, aiming to satisfy both consumers and farmers, often suppress the profit margin and demand that FMBs reduce their expenses. While the government typically has the final say, any deficit incurred is ultimately borne by them, rendering the victory somewhat hollow.

The release of stocks is largely influenced by consumer needs. Since staple crops typically experience steady demand throughout the year, FMBs must manage the significant gap between post-harvest purchasing and the gradual sale of goods over time. Stockholding is crucial but costly, especially immediately after harvest when storage space for the incoming produce is often insufficient. Conversely, FMBs tend to operate with underused storage for much of the year as stocks are slowly released. A central goal for many FMBs is ensuring "food security" during times of shortage. This policy makes political sense but presents commercial challenges. It requires prolonged working capital, and if no shortage occurs, FMBs are left with deteriorating stocks.

In almost all regions, there is a "dual marketing system," with a parallel market that allows farmers, traders, and consumers to bypass the FMB. In some countries, the government permits this parallel market, while in others, where FMBs hold a monopoly, it becomes an illegal black market, the suppression of which has often proven impossible. Whether the parallel market is legal or not, FMBs must contend with its competition.

Some marketing boards do not neatly fit into either of the two categories described above. For products like groundnuts, sunflower seed oil, and palm oil, which have both domestic and export markets, the marketing boards handling these items were primarily established in countries with an export surplus. These boards are typically categorized as export boards. However, if domestic demand rises to the point of absorbing the export surplus, the board may shift to being a domestic marketing board.

When evaluating the economic performance of marketing boards, it is crucial to recognize that their objectives often extend far beyond commercial aims and involve fulfilling political goals. In fact, policymakers frequently refer to marketing boards as "instruments" for achieving broader national objectives. Those who quickly point out the "mistakes" of marketing boards often overlook the need to differentiate between errors that can be directly attributed to the management of the boards, which can be corrected, and those that stem from external influences.

Many of the criticisms would be better directed at those overseeing the boards' activities. For instance, marketing boards are often used by governments as tools for national policy, which include:

- Promoting agricultural and rural development, where social goals take precedence over commercial ones
- Serving as instruments of fiscal policy
- Controlling urban wages by limiting the prices of staple foods
- Encouraging farmers to increase production of food and export crops by offering higher producer prices
- Strengthening political power through the appointment of political allies to the board

The operational challenges and decisions that adversely affect the profitability and cash flow of marketing boards are often not the result of poor commercial decision-making by management. Instead, they can be traced to individuals outside the boards themselves, whose objectives are disconnected from the effective and efficient functioning of the boards.

Co-operatives in the Agriculture and Food Sectors

Overall, the portrayal of marketing boards in the literature is largely disheartening, with many being depicted as ineffective organizations that have failed to achieve significant success. Because they empower farmers, increase market accessibility, and guarantee fair prices for agricultural products, cooperatives are essential to India's food and agriculture industries. The foundation of these organizations is the idea of collective effort, in which small producers and farmers band together to create cooperatives in order to increase their negotiating power, lower production costs, and gain access to better financial and marketing resources. Cooperative movements have a long history in India and are still a major factor in the economic growth of rural areas.

When evaluating the economic performance of marketing boards, it is crucial to recognize that their objectives often extend far beyond commercial aims and involve fulfilling political goals. In fact, policymakers frequently refer to marketing boards as "instruments" for achieving broader national objectives. Those who quickly point out the "mistakes" of marketing boards often overlook the need to differentiate between errors that can be directly attributed to the management of the boards, which can be corrected, and those that stem from external influences. Many of the criticisms would be better directed at those overseeing the boards' activities. For instance, marketing boards are often used by governments as tools for national policy, which include:

- Promoting agricultural and rural development, where social goals take precedence over commercial ones
- Serving as instruments of fiscal policy
- Controlling urban wages by limiting the prices of staple foods

- Encouraging farmers to increase production of food and export crops by offering higher producer prices
- Strengthening political power through the appointment of political allies to the board

The operational challenges and decisions that adversely affect the profitability and cash flow of marketing boards are often not the result of poor commercial decision-making by management. Instead, they can be traced to individuals outside the boards themselves, whose objectives are disconnected from the effective and efficient functioning of the boards.

Types of Agricultural and Food Cooperatives in India

Dairy Cooperatives

In India, dairy cooperatives are among the most prosperous cooperative models. They have played a major role in India becoming the world's largest milk producer.

- **Amul (Gujarat Cooperative Milk Marketing Federation - GCMMF):** Founded in 1946, Amul revolutionized the dairy industry through the White Revolution, ensuring fair prices for dairy farmers and providing quality milk and milk products nationwide.
- **Nandini (Karnataka Milk Federation):** A leading dairy cooperative in Karnataka that supports thousands of dairy farmers with procurement, processing, and marketing of milk products.
- **Mother Dairy:** Established by the National Dairy Development Board (NDDB), it plays a vital role in the dairy supply chain, ensuring the availability of fresh milk and dairy products.

Agricultural Credit Cooperatives

Credit cooperatives provide essential financial support to farmers who struggle to obtain loans from commercial banks. They help in offering low-interest loans for purchasing seeds, fertilizers, and equipment.

- Primary Agricultural Credit Societies (PACS): These are grassroots-level cooperatives that provide short-term credit to farmers.
- State Cooperative Banks (SCBs) and District Central Cooperative Banks (DCCBs): They form an essential part of India's rural credit structure, offering financial services to small farmers and agricultural workers.
- National Cooperative Development Corporation (NCDC): This institution supports cooperatives in various sectors, including agriculture, with funding and development programs.

Agricultural Marketing Cooperatives

These cooperatives help farmers market their produce efficiently, eliminating the role of middlemen and ensuring better price realization.

- **National Agricultural Cooperative Marketing Federation of India (NAFED):** Established in 1958, NAFED assists farmers in selling their produce at fair prices, stabilizing agricultural markets.
- **State Agricultural Marketing Cooperatives:** These organizations manage procurement, storage, and distribution at the state level, helping to maintain market stability.

Sugar Cooperatives

India is one of the largest producers of sugar, and cooperative sugar mills play a crucial role in the industry.

- **Maharashtra Sugar Cooperatives:** Maharashtra has a significant number of sugar mills run by cooperatives, supporting sugarcane farmers and boosting rural employment.
- **Uttar Pradesh Cooperative Sugar Mills:** These mills provide financial and technical assistance to sugarcane growers in the state.

Fertilizer Cooperatives

Fertilizer cooperatives ensure the availability of quality fertilizers to farmers at affordable prices.

- **Indian Farmers Fertilizer Cooperative Limited (IFFCO):** One of the world's largest cooperatives, IFFCO provides fertilizers and agronomic services to millions of farmers.
- **Krishak Bharati Cooperative Limited (KRIBHCO):** Another leading cooperative that produces and distributes fertilizers, enhancing soil fertility and productivity.

Food Processing Cooperatives

These cooperatives add value to agricultural produce by processing raw materials into finished food products.

- **Horticultural Producers' Cooperative Marketing and Processing Society (HOPCOMS):** Based in Karnataka, HOPCOMS helps farmers in the marketing and processing of fruits and vegetables.
- **Lizzat Papad:** A famous women's cooperative that manufactures and markets high-quality food products, providing employment to thousands of rural women.

The Structure and Organisation of Co-operatives

There are two principal forms of co-operative organisations: primary co-operatives and secondary co-operatives. The basic unit in the co-operative systems is the primary co-operative. A primary co-operative is one in which the shareholder are individuals; each of them having an equal share in its control.

Primary co-operatives	In many cases, primary co-operatives will combine several functions e.g. an agricultural co-operative may provide consumer supplies to its members. Primary co-operatives may also own and run subsidiary enterprises related to their main functions, such as a consumer co-operative with its own manufacturing/processing or servicing business.
Secondary co-operatives	While a primary co-operative has individual persons as members, a secondary (or federal) co-operative is one in which other co-operatives are the members. Apart from this basic difference the structure and organisation of both types follow a very similar pattern.

The Control and Management of Primary Co-Operatives

The governance structure of cooperatives consists of three levels, as shown in Figure. The General Meeting of Members is responsible for setting policy, and through this meeting, members exert control over the cooperative. In many countries, it is legally required to hold an Annual General Meeting (AGM), which has specific duties, including reviewing and approving an audited financial statement, deciding how any surplus will be allocated and distributed, and electing committee members.

The General Meeting of Members delegates the operational control of the co-operative to a management committee (or board of directors), which controls the works of the co-operative on behalf of the members. One member of this committee is elected chairman or president. A manager (or secretary) is appointed by the management committee as the chief administrative officer of the co-operative. He/she is responsible to the committee for the day-to-day control of the business. In small co-operatives he/she may be a member elected to do the work without pay.

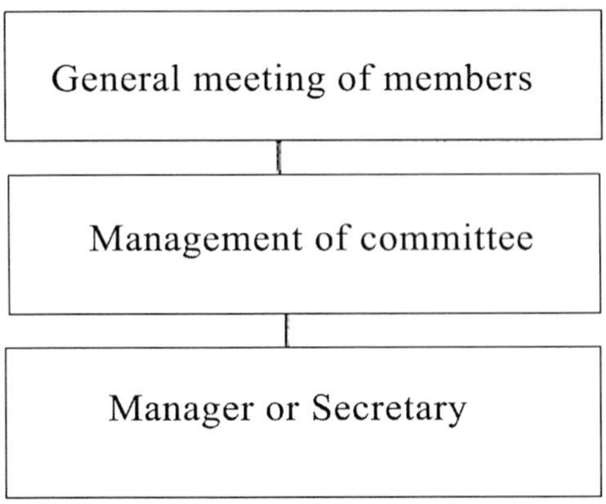

Figure: 7.7 The Management Structure of Primary Co-Operatives

Federal or Secondary Co-Operatives

Secondary co-operatives, often referred to as "union" or "federal" co-operatives, can be established for a variety of purposes. It is common for a primary co-operative to be a member of multiple secondary co-operatives, depending on its needs and the structure of the local co-operative system. Examples of secondary co-operative organizations include:

- A local district union of 3 or 4 cotton marketing co-operatives operating a ginning plant
- A federation of two or more consumer co-operatives running a bakery
- A national union of agricultural co-operatives manufacturing fertilizer
- A national union of consumer co-operatives managing wholesaling services and producing merchandise
- A national federation of co-operatives managing a national bank

Through the process of federation, co-operatives can organize large-scale operations at national or even international levels without compromising the democratic control of the primary co-operatives by their members. The secondary co-operative, due to its larger business volume or broader representation, can carry out functions, provide services, and make decisions that would be beyond the capacity of most primary co-operatives. Secondary co-operatives are a form of vertical integration that offers opportunities for economies of scale, development, and better administration.

These secondary co-operatives can also form additional layers of secondary co-operatives, known as tertiary co-operatives. In many countries, there is an apex federation that represents all co-operatives, offering national-level advisory, representative, and professional services to the co-operative movement. These national federations can also affiliate with international organizations like the International Co-operative Alliance.

Control and Management of Secondary Co-Operatives

The control and management structure of secondary co-operatives is similar to that of primary co-operatives. The shareholding members, which are the primary co-operatives, exercise policy control through the General Meeting and elect a management committee to act on their behalf. This committee then appoints a chief officer to manage operations under its guidance. The by-laws of secondary co-operatives, just like those of primary co-operatives, outline the organizational rules and procedures and must be approved by the relevant local authority. The operating surplus of a secondary co-operative is used and distributed according to the same principles as those followed by primary co-operatives.

However, there is a risk that federal co-operatives could become the dominant force over their member co-operatives, rather than serving them. This can happen for two reasons. First, secondary co-operatives engaged in manufacturing and trade generally require a high level of integration between their operations and those of their members to operate efficiently. This often necessitates a high degree of discipline or some form of contractual obligation. Secondly, the size and trade volume of secondary co-operatives can be so large, compared to individual primary co-operatives, that they may begin to assume a dominant role in the relationship. This tendency must be monitored and, if necessary, controlled. Ultimately, the primary responsibility of the secondary co-operative is to provide services to its member co-operatives.

The potential of co-operatives is vast, and they are well-suited to meet the economic, social, and institutional needs of rural development. Co-operatives offer a means to organize and mobilize people for self-help actions, enabling them to access the services they require as farming and rural communities. As self-managed rural institutions, co-operatives can adapt to and respond to the needs of their members while fostering self-reliance and self-confidence within a shared vision and cooperative efforts. In providing services to their farmer-members, co-operatives play a vital role in supporting both the farmers' development objectives and broader national development goals.

As business entities, co-operatives are capable of acting as development agencies in their own right. Through the accumulation of business assets, the expansion of services, the development of management expertise, and the creation of employment, they contribute directly to positive and measurable development outcomes. Furthermore, the adaptable nature of co-operative structures, such as secondary co-operatives, offers opportunities for collective action in the growth of agro-industrial enterprises, which can support and enhance local initiatives, providing additional momentum for rural development. Co-operatives are widely recognized in many countries as powerful drivers of development.

Cooperatives Shortcomings

The potential of co-operatives has, in many cases, not been fully realized, and their development has often fallen short of expectations. Low performance standards, poor management, financial failure, corruption, misuse of funds, and political manipulation have all been recurring issues in many co-operative enterprises. As a result, co-operatives have faced significant criticism, with some members even becoming cynical about their ability to contribute effectively to the development process. There are several key challenges that hinder co-operative growth and impact their performance, which are discussed below:

- **Realism of Objectives:** Commitment and purpose are essential for motivation, and achieving objectives is equally important. Objectives represent the purpose and expectations, and to motivate and guide actions, they must be attainable. Adequate resources must be available to achieve the objectives, and aspirations must align with capability. Co-operatives and their members should not expect too much from them, especially not rapid expansion. Agricultural co-operatives in developing countries often operate in conditions that any business would find challenging. Like their members, co-operatives function in difficult circumstances, with high operating costs, low margins, limited turnover, narrow stock inventories, and exposure to the impacts of crop failure. These challenges are common for businesses operating in such environments, and if they weren't present, private enterprise would likely take over a profitable market. Co-operatives are often established to provide essential services in areas where other agencies have failed or refused to do so.

Expecting too much, too quickly, can also be problematic. It's a common mistake to assume that once a co-operative is reasonably established, an influx of external loan capital will enable rapid growth. However, such sudden investments can overwhelm management resources, encourage risky behavior, compromise financial judgment, lead to excess inventory, and result in loss-making ventures. Co-operatives should be allowed to grow at a pace matching their members' ability to manage, control, and finance the development. Expansion should be gradual, supported by the surpluses from their own operations, rather than driven by external government policy goals. Sudden, revolutionary growth can harm the viability of the co-operative and hinder the achievement of long-term objectives.

- **Conflict Between Economic and Social Purposes:** Economic success is essential for achieving co-operative goals because, in the long term, unprofitable enterprises cannot survive. However, co-operatives face limitations in replicating capitalist business models without compromising the fundamental values of the co-operative movement. For instance, pursuing business growth can tempt management to weaken member control, focus solely on profit, and disregard equity and fairness. The reinvestment of profits to build collectively-owned capital is a desirable practice but can have drawbacks. If member share capital becomes too small in relation to the total capital structure, professional management might disregard member control, making decisions without considering members' interests. In such cases, the co-operative becomes largely indistinguishable from a capitalist enterprise, despite retaining its name.
- **Misuse of Co-operatives for Political Purposes:** When the purpose and resources of co-operatives are diverted for political gain, it negatively impacts the co-operative movement. Political divisions among members can distract from economic goals, turning meetings into political battlegrounds. Discontent among members can arise, making it easy for a minority to seize control and direct the co-operative for their own political ends. Co-operative principles dictate that membership should not imply political commitment or obligation. Co-operatives tightly controlled by governments as instruments of economic policy are rarely conducive to democratic, member-owned control. They are often created to serve political agendas, which may not align with the interests of the members, who have limited control over the enterprise.
- **Management:** There has been a tendency to argue that a primary reason for co-operative failure is the limitation placed on management skills and authority due to the democratic structure of the enterprise. Consequently, some suggest that the authority of the General Meeting should be reduced, allowing committees and managers to focus solely on managing the operations. However, implementing such a change would undermine the core purpose of the co-operative, which is to empower people to run their own business. Instead

of restricting member participation, Furthermore, the solution should be to enhance and improve their involvement. the management standards within co-operatives are often subpar. As previously mentioned, co-operatives are often established in markets and areas that are considered marginal in terms of profitability by other business enterprises. As a result, the salaries, working conditions, and locations they offer fail to attract high-quality managers.

Co-Operatives and Their Members Selling Arrangements

A principal policy question in co-operatives is the procedure to be used in selling members' produce. The alternatives are: outright purchase from members, or sale on commission.

- **Outright Purchase:** In this system, members are paid for their produce based on prices set by the co-operative at the time of delivery, and the co-operative takes ownership of the produce. It then resells the produce under the best possible terms it can secure. The profits generated from these transactions are first used to cover operating expenses, with any remaining surplus being allocated or distributed according to the decision of the General Meeting. This method requires the co-operative to maintain significant levels of funding. Since, in the case of seasonal crops, a large amount of produce is typically available immediately after harvest, this can result in a serious cash flow issue. One solution to this is a two-stage payment system, where members receive part of the sale price at the time of delivery and the remainder after the co-operative has resold the produce.

  The main downside of outright purchase is that the co-operative assumes all the post-harvest risks, which include: a decrease in demand, price fluctuations, loss of value due to product downgrading, deterioration leading to a decline in quality and value, transportation failures, spoilage, fire, and theft. While some of these risks can be covered by insurance, most cannot. This approach is generally only viable when the risks are manageable and can be reasonably estimated, such as when forward contracts have been secured. Despite these risks, the advantage of outright purchase is that it allows the co-operative to add value to the crop, thereby increasing the profits of its members. There are three main ways in which a co-operative can add value to a commodity:

- Produce can be stored and sold later when prices are higher.
- Primary processing of the crop can add value. For example, cotton ginning is a simple and cost-effective process that is best done near the fields. The value of ginned cotton is typically much higher than the value of raw cotton plus the ginning costs, and there is additional value in the cotton seeds.
- Value can be added through the packaging and presentation of the crop, or by improving the quality or condition of livestock before sale. Graded, washed, and well-packed fruits and vegetables, for instance, can attract broader markets and higher prices.

However, it should be noted that while price advantages gained from adding value directly benefit both the co-operative and its members, the additional investment required to achieve these returns can be prohibitive. In addition to the risks involved, outright purchase, storage, processing, packaging, transportation, and marketing all require substantial financing. Furthermore, these more complex operations demand a high level of management expertise and judgment, which is often in short supply.

- **Sale on commission:** This much simpler, virtually risk-free system positions the co-operative as an agent for the producers without assuming legal ownership of the goods. As a result, all risks remain with the individual producers. The co-operative collects produce from its members and sells it in the best available markets, deducting a commission at a previously agreed-upon rate from the sale price. The co-operative covers its expenses through the income from these commissions. By using the sale-on-commission method, the co-operative avoids the need to finance the purchase of crops and minimizes its exposure to risk. Additionally, the operation requires simpler procedures, and expenditure can be more accurately aligned with expected income.

The primary disadvantage of the sale-on-commission approach is that neither the co-operative nor the member can benefit from potential price increases. Another issue is the delay in the producer receiving payment for their crops. The co-operative cannot make payments until it has been paid by the customer. In addition to the time required for crops to be sold in an open market, it is not uncommon for state-run agencies to be slow in disbursing payments for produce received from co-operatives.

IMPORTANCE OF COOPERATIVES IN AGRICULTURE AND FOOD SECTORS

Cooperatives in agriculture and food sectors help small and marginal farmers overcome challenges such as market volatility, lack of credit facilities, and exploitation by intermediaries. They enable farmers to collectively purchase inputs, process agricultural products, and distribute them efficiently. Moreover, cooperatives contribute to rural employment, improve food security, and ensure a stable supply of essential food commodities.

Challenges Faced by Agricultural and Food Cooperatives

Despite their success, cooperatives in India face several challenges:

- **Lack of Professional Management:** Many cooperatives suffer from inefficient administration and decision-making.
- **Political Interference:** Some cooperatives are heavily influenced by political interests, affecting their operations and objectives.
- **Limited Access to Technology:** Many cooperatives lack modern technology and digital infrastructure for efficient marketing and processing.

- **Financial Constraints:** Some cooperatives struggle with financial mismanagement, leading to losses and reduced benefits for members.

Government Initiatives to Strengthen Cooperatives

To address these challenges, the Indian government has launched several initiatives to strengthen the cooperative movement:

- **e-NAM (National Agriculture Market):** A digital marketplace that helps cooperatives and farmers access better markets and fair pricing.
- **Formation of Farmer Producer Organizations (FPOs):** Encourages small farmers to come together as producer organizations, similar to cooperatives, for better market access.
- **Support from NABARD:** The National Bank for Agriculture and Rural Development (NABARD) provides financial and policy support to cooperatives.

Future Prospects of Cooperatives in India

With technological advancements, policy support, and increasing consumer demand for organic and locally sourced products, agricultural and food cooperatives have a bright future in India. By adopting modern management practices, improving financial transparency, and leveraging digital marketing, cooperatives can continue to be a driving force for India's rural and agricultural development.

Cooperatives in the agriculture and food sectors of India have played a transformative role in empowering farmers, improving productivity, and ensuring food security. While challenges remain, with continued support and modernization, cooperatives can further strengthen India's agricultural economy, benefiting millions of farmers and consumers alike.

8

Warehouse Management

Warehouse management plays a pivotal role in supply chain operations, concentrating on the effective storage, organization, and handling of goods within a warehouse. This encompasses a variety of tasks such as inventory control, order processing, and ensuring that products, especially those that are perishable or sensitive like food, pharmaceuticals, and electronics, are stored under proper conditions. Well-managed warehouses guarantee that goods are stored in environments that protect them from spoilage, damage, or contamination, while also allowing for fast retrieval when needed for distribution.

Key elements of warehouse management include the design of the warehouse layout, inventory tracking systems, automation, and workforce management, all working together to boost efficiency. As global supply chains become more complex, warehouse management has adapted by incorporating advanced technologies like RFID, warehouse management systems (WMS), and automation, all of which help streamline operations, cut costs, and enhance overall productivity. Efficient warehouse management is critical in ensuring the integrity of the supply chain, boosting customer satisfaction, and making sure that products reach their destination on time and in perfect condition.

WAREHOUSING: THE BASICS

Warehouses are specially designed, scientifically constructed storage facilities built to safeguard the quantity and quality of stored goods. In order to guarantee the availability and safety of goods until they are sold or distributed, warehousing is the process of managing and keeping them in a specific location. Order fulfillment, inventory management, receiving, storing, and quality control are all part of it. When it comes to retail establishments or online businesses, this kind of logistics is essential because it guarantees prompt delivery for the satisfaction of customers and ensures that there are sufficient resources to uphold staff supervision and equipment safety regulations.

The purpose of warehouse space extends beyond basic stocking; it is to support operational excellence by effectively managing movements across the whole system involved in these activities, whether internally (within-house) or externally (with third parties).

THE ELEMENTS OF WAREHOUSING

While warehousing may appear straightforward, as it primarily involves storing products, several critical processes must be followed to ensure efficiency and facilitate the smooth movement of inventory. These include:

- **Capacity Planning:** Since space is the primary resource in warehousing, staff must plan in advance where incoming shipments will be stored to maximize space utilization and maintain operational efficiency.
- **Receiving Inbound Shipments:** Upon arrival at the warehouse, products must be carefully received and transferred to a staging area for further processing. This step ensures that items are properly accounted for and prepared for storage.
- **Tracking Inventory:** As products move in and out of the facility, they must be logged into the warehouse inventory management system. This allows administrators to monitor current stock levels, track product movement, and plan for future inventory adjustments.
- **Storing Products:** Once received and processed, products need to be placed in designated storage areas. This may involve organizing items in bins or on pallets and using handling equipment to transport them to their assigned locations within the warehouse.
- **Climate Control:** Certain products require specific environmental conditions, such as controlled temperature, humidity, or pressure. For instance, frozen goods must be kept in areas where the temperature remains below freezing. These requirements influence storage decisions to maintain product integrity and quality.
- **Reorganizing Inventory:** As new shipments arrive, existing stock may need to be relocated to optimize space utilization. Any changes made in storage locations must be updated in the inventory management system to ensure accurate tracking.
- **Retrieving and Outbound Shipping:** When products are scheduled for dispatch, warehouse staff must retrieve, process, package, and load them for shipment. These items are then marked as released from inventory, creating space for new incoming shipments and maintaining warehouse efficiency.

Need of a Warehouse

Although, there are several justifications for owning a warehouse, the primary ones are:

- **Stockpiling:** When seasonality is present, the warehouse's role as a buffer between supply and demand becomes even more crucial.
- **Production:** In order to achieve economies in the production process, warehousing is required for the storage of components, sub-assemblies, work-in-progress, finished goods, and raw materials.

- **Product mixing:** Product mixing is the process of combining multiple products into a single product through packaging.
- **Consolidation:** The warehouse serves as a hub for the "pulling" of goods from suppliers into a location for sorting before sending them on to the end user. This warehouse can only be used for sorting; no stock is kept there.
- **Distribution:** Goods can be "pushed" from the manufacturer to the final consumer via the warehouse network.
- **Customer service:** Delivery time is the primary service level that the warehouse is there to provide.

Working of Warehouses

- **Acts:** The operations of warehouses, such as those managed by the Central Warehousing Corporation (CWC) and State Warehousing Corporations (SWCs), are governed by the relevant Warehousing Acts enacted by either the Central or State Government.
- **Eligibility:** Any individual may store designated commodities in a warehouse, provided they agree to pay the stipulated charges.
- **Warehouse Receipt (Warrant):** This is a document issued by the warehouse manager or owner to the person storing their goods. The receipt includes details such as the warehouse's name and location, the issuance date, and a description of the commodities, including the grade, weight, and estimated value based on current market prices.
- **Use of Chemicals:** To maintain the quality of the produce, the goods stored in the warehouse are preserved using scientific methods to protect them from rodents, insects, pests, and other types of infestations. The warehouse bears the cost of periodic dusting and fumigation to ensure the produce remains in good condition.
- **Financing:** The warehouse receipt acts as collateral, enabling the holder to secure credit.
- **Delivery of Produce:** In order to withdraw goods, the warehouse receipt must be presented to the warehouse owner. The holder may also take possession of a portion of the stored goods after paying the applicable storage fees.

Importance

Scientific Storage

These structures protect products from both quantity and quality loss by employing appropriate preservation techniques tailored to the nature of the stored items.

Financing

Warehouses address the financial requirements of individuals storing their goods. Nationalized banks provide credit secured by the warehouse receipt issued for the stored products, often advancing 75 to 80% of the product's value.

Price Stabilization

By reducing the pressure to sell immediately after harvest, warehouses contribute to the stabilization of prices for agricultural commodities.

Market Intelligence

Warehouses provide holders of stored produce with valuable market information, enhancing their ability to make informed decisions regarding the sale or distribution of their goods.

TYPES OF WAREHOUSES

Based on Ownership

- **Private Warehouses:** These are owned by individuals, large corporations, or wholesalers who use them for storing their own products. They also offer storage services to third-party clients.
- **Public Warehouses:** These warehouses are owned by the government and are primarily used for the storage of goods, serving the public's storage needs.
- **Bonded Warehouses:** Located typically at seaports or airports, these warehouses are designed to store imported goods until the customs duties are paid by the importer. Bonded warehouses are licensed by the government for this specific purpose, and the goods stored are considered bonded. The services offered by bonded warehouses include: i. Allowing the importer to avoid paying all customs duties at once by enabling the delivery of goods in multiple installments. ii. Regular maintenance operations like spraying and dusting to preserve the quality of the goods. iii. Facilitating entrepot trade (re-export of goods) by offering storage for goods intended for re-export.

Based on the Type of Commodities Stored

- **General Warehouses:** These are standard warehouses used for storing a wide variety of goods, including food grains, fertilizers, and other general products.
- **Special Commodity Warehouses:** These warehouses are specifically designed and constructed to store certain types of commodities, such as cotton, tobacco, wool, and petroleum products.
- **Refrigerated Warehouses:** These warehouses maintain specific temperatures to cater to the storage needs of perishable goods such as vegetables, fruits, fish, eggs, and meat.

STORAGE FACILITIES INDIA

Central Warehousing Corporation (CWC)

Established as a statutory body on 2nd March 1957 in New Delhi, the Central Warehousing Corporation offers secure and dependable storage services for around 120 agricultural and industrial goods.

Functions of CWC

- To acquire land and construct warehouses and godowns at appropriate locations throughout India.
- To operate warehouses for the storage of agricultural products, seeds, fertilizers, and other designated goods on behalf of individuals, cooperatives, and various institutions.
- To act as the government's agent for the procurement, storage, sale, and distribution of these goods.
- To facilitate the transportation of these commodities.
- To invest in the share capital of State Warehousing Corporations.
- To perform any other duties as prescribed under the relevant Act.
- The CWC operates air-conditioned godowns in cities such as Calcutta, Bombay, and Delhi, and also provides cold storage facilities in Hyderabad.
- Special storage methods are offered by CWC to preserve delicate and hygroscopic products.
- Additionally, the corporation has developed specialized storage techniques for commodities like spices, coffee, seeds, and more.

State Warehousing Corporations (SWCs)

In various states of India, individual warehousing corporations have been established to cater to storage needs. These corporations primarily operate in district-level centers. The share capital of each State Warehousing Corporation is jointly funded by the respective state government and the Central Warehousing Corporation.

Food Corporation of India (FCI)

In addition to the CWC and SWCs, the Food Corporation of India (FCI) has also developed substantial storage infrastructure. FCI stands as the largest agency in the country, with a storage capacity of 26.62 million tonnes.

PRELIMINARY OPERATIONS IN WAREHOUSE

Presizing

For many commodities, fruit below a certain size is eliminated either manually or mechanically by a presizing belt or chain. Undersized fruit is diverted to a culling conveyor or used for processing.

Sorting or Selection

The sorting process eliminates cull, overripe, misshapen, and otherwise defective fruit and separates products by colour, maturity, and ripeness classes (e.g., tomato and muskmelons).

Grading

Fruit are sorted by quality and placed in separate grades. There are different international grading standards for many commodities, but these are not always used.

Waxing

Food-grade waxes are typically applied to vegetables like cucumbers, eggplants, sweet peppers, rock melons, and tomatoes, and occasionally to certain summer squashes. This process aims to replace some of the natural waxes that are lost during cleaning and washing, helping to prevent water loss and improve the appearance of the produce. Waxing can take place either before or after the sizing process, and fungicides may be added to the wax. It's important to note that both the application of wax and post-harvest fungicides must be clearly marked on each shipping container. Waxing and fungicides are applied exclusively during the packinghouse handling of fruit vegetables.

Sizing

Once the fruit vegetables are sorted to remove defects and color inconsistencies, they are divided into various size categories. Sizing is performed manually on many types of fruit vegetables, such as legumes, soft and hard rind squashes, cucumbers, eggplants, capsicums, okras, pumpkins, muskmelons, and watermelons. For example, rock melons may be sized using volumetric weights or diverging roll sizers, capsicums are typically sized with diverging bar sizers, and tomatoes are sized either by diameter using belt sizers or by weight.

Packing

The packing process involves placing the sorted and sized produce into containers of predetermined sizes, such as cartons, flats, or barrels. Some crops benefit from specialized mechanical harvesters, which also carry out preliminary tasks such as presizing, sorting, and packing. Although, these packers usually come with a higher initial investment, they have lower operating costs and can pay for themselves within a few years.

For other fruit-type vegetables, packing is often done by hand or mechanically into shipping containers, using methods like count-based packing, bulk bins (for hard rind squashes, pumpkins, muskmelons, and watermelons), or bulk trucks (for watermelons). Fruit vegetables that are place-packed are often sized during the same packing operation.

Palletising

In large-volume operations, packed shipping containers of fruit vegetables are palletised for transport. This practice is common with produce such as rock melons, muskmelons, capsicums, and tomatoes. For immature fruit vegetables, such as sweet corn, palletising is less common in low-volume operations due to the lack of forklifts. In these cases, palletising typically occurs at a centralized cooling facility or when products are loaded for transport. Palletising is usually done after hydrocooling or package-ice cooling but before forced-air cooling. In field-packing operations, palletising is generally completed in the field itself.

Cooling

Cooling methods for fruit vegetables vary depending on the type of produce. Forced-air cooling is a common technique used for beans, rock melons, cucumbers, muskmelons, peas, capsicums, soft-rind squashes, and tomatoes. For chilling-sensitive commodities, such as squashes, peppers, eggplants, and cherry tomatoes, forced-air evaporative cooling is applied selectively. Hydrocooling is frequently used before grading, sizing, and packing for items like beans, cantaloupe, sweet corn, and okra. Sorting for defective products occurs both before and after the cooling process, but during hot weather, hydrocooling cycles may not be enough. In such cases, allowing additional time in cold rooms after packing and palletising ensures proper cooling before transport.

Package icing and liquid-icing methods are employed to cool rock melons and supplement hydrocooling for sweet corn. In large-scale operations, temporary cold storage is utilized, with fruit vegetables placed in cold storage rooms after cooling and before shipment. In smaller farm operations, products are typically transported to central cooperatively owned or distributor-owned facilities for cooling and short-term storage. When it comes to transportation, some tomatoes, rock melons, and muskmelons are shipped in refrigerated railcars, but most fruit vegetables are transported in refrigerated trucks or container vans. While some products like cantaloupe and tomatoes may be shipped together, it is crucial to separate ethylene-sensitive commodities. Immature fruit vegetables such as cucumbers, legumes, bitter melons, and eggplants are sensitive to ethylene exposure, and for mature fruit types, watermelon can be negatively affected by ethylene, resulting in softening, mealiness, and rind separation.

THE IMPORTANCE OF WAREHOUSE MANAGEMENT

Operating a warehouse is primarily a cost-adding activity rather than a value-adding one. It could be argued that you are not adding value if you are holding materials or goods without altering them or bringing them closer to your client. The cost of operating and maintaining a warehouse, in addition to the initial building and equipment costs, is incurred when materials or goods are stored there.

Therefore, it is crucial that the operation be conducted as cost-effectively as possible and that the warehouse activity add value wherever it can. According to logistics cost analysis, warehouse expenses make up around 30% of the total. It goes without saying that this will change depending on the industry sector. The type of product will determine a lot of things, such as how much room each item requires and how it is handled, stacked, and how much stock it can hold.

In the supply chain process, goods in all of their forms must be stored at different points in time. However, it is crucial that they move through the warehouse as fast as possible to avoid incurring extra costs. The speed at which goods can be moved through the warehouse will be significantly impacted by how it is designed, organized, and run.

When items are in a warehouse, it's critical to use the operation to add value wherever possible in addition to focusing on efficiency, minimizing expenses, and cutting back on non-value-added activities. Consolidating goods, selecting and assembling orders, and customizing are a few examples of value-added activities. You haven't added any value to the goods if you bring them into a warehouse, store them there, and then send them out in the same condition.

9

Cold Storage Management

India stands as the largest global producer of fruits and the second-largest producer of vegetables. Despite this, the per capita availability of fruits and vegetables remains relatively low, primarily due to post-harvest losses that account for approximately 25% to 30% of total production. Furthermore, the quality of a significant portion of produce deteriorates by the time it reaches consumers. A major challenge in the marketing of fruits and vegetables stems from their perishable nature, which leads to high marketing costs, market surpluses, price fluctuations, and similar issues. However, when stored at low temperatures, the perishability of produce is substantially reduced, and its shelf life is extended, highlighting the importance of cold storage and refrigeration. The first cold storage in India was established in Calcutta in 1892, but the expansion of the cold storage industry gained substantial momentum only after the country's independence. To ensure proper conditions in cold stores and promote the scientific development of the industry, the Government of India and the Ministry of Agriculture introduced the "Cold Storage Order, 1964" under Section 3 of the Essential Commodities Act, 1955. The Agricultural Marketing Advisor to the Government of India acts as the Licensing Officer. Accessible cold storage facilities can significantly reduce the risk of distress sales and ensure better returns for producers.

Fresh food products, especially fruits and vegetables, are highly perishable and susceptible to pests and spoilage. However, controlling temperature and relative humidity, along with preventing damage, can greatly enhance their shelf life. Biological deterioration, caused by factors like respiration rate, ethylene production, mechanical injuries, water stress, physiological disorders, and pathological breakdowns, leads to decay, a loss of nutritional value, and changes in color, texture, and flavor. The rate of deterioration is influenced by factors such as temperature, relative humidity, air velocity, the atmospheric composition (levels of oxygen, carbon dioxide, and ethylene), and sanitation. Mold, pests, and other spoilage causes that lead to quality deterioration, such as water loss, thrive in warm, humid conditions—conditions that are common in India. Most factors that cause product deterioration can be mitigated through temperature control, primarily cooling. Reducing temperature slows down physiological processes (like respiration, ethylene production, and enzymatic reactions) and inhibits the growth of microorganisms (bacteria and fungi). After harvest, the produce remains "alive," meaning metabolism processes like respiration and microbial activity continue, leading to deterioration. To lower respiration rates, delay ripening, and reduce water loss, temperature control is essential—either by avoiding heat exposure or by creating cooling conditions, which extends shelf life. Adequate storage facilities help control factors like temperature, relative humidity, and air velocity, enhancing product quality, shelf life, and value.

The table below illustrates how temperature control affects the shelf life of fresh produce. Reducing temperature, even if not at optimal storage conditions, can significantly extend storage life. Consequently, any cooling method, such as placing produce in shaded, ventilated areas or selecting early morning or evening hours for harvesting, offers substantial benefits.

Table 9.1 Effect of temperature control on shelf-life of fresh produce

Temperature (°C)	Relative Rate of Deterioration	Relative Shelf Life	Example of Potential Shelf Life in Days
0	1.0	100	45
10	3.0	33	15
20	7.5	13	5
30	15.0	7	2.5
40	22.5	4	1.3

Table 9.2 Effect of temperature control on shelf-life of fresh agricultural products

Commodity	Recommended Temperature for Handling and Storage (max post-harvest life)	Post-harvest Life at 35 °C (ambient temperature)	Post-harvest Life at 25 °C	Post-harvest Life at 15 °C	Increased Marketing Time Available at 15 °C
Cabbage	0°C (6 month)	2 weeks	4 weeks	8 weeks	4X
Carrots	0°C (6 month)	2 weeks	4 weeks	8 weeks	4X
Tomatoes	15°C (14 days)	3 days	6 days	14 days	5X
Peppers	12°C (20 days)	3 days	7 days	15 days	5X
Potatoes	5°C to 7 °C (5 to 10 months)	2 weeks	4 weeks	8 to 10 weeks	4X
Spinach	0°C (14 days)	1 day	2 months	5 days	5X

Once produce has been pre-cooled, it must be transported and placed into a cold storage facility where it should be kept at its optimal storage temperature to maximize shelf life and maintain quality. This optimal temperature varies depending on the type of produce, which can create challenges if multiple items need to be stored in the same facility, especially when different storage chambers are not available. The storage conditions, alongside the intrinsic characteristics of the produce, significantly influence its shelf life.

In general, the storage temperature should be as low as possible while still being above the threshold at which the specific crop becomes sensitive to chilling. Another critical factor affecting the shelf life of fresh produce is the level of relative humidity, which impacts water loss, decay, physiological disorders, and even the uniformity of ripening. For most produce, high relative humidity (saturated levels) is ideal. Furthermore, ethylene plays a key role in the shelf life of fresh produce. While some fruits and vegetables are ethylene producers after harvest, others are sensitive to ethylene exposure. Therefore, if multiple types of produce must be stored together, their individual requirements for temperature, humidity, and ethylene production or sensitivity must be taken into account.

The Indian National Board for Horticulture has divided produce into 7 categories according to optimum storage temperature and relative humidity.

COLD STORAGE TECHNOLOGIES FOR SMALL-SCALE PRODUCTION

Cooling and cold storage are highly energy-intensive and costly processes that rely heavily on a reliable electricity supply. However, in many developing countries, such a supply is often unavailable, and the capital and operating expenses for farmers are typically too high. As a result, small-scale farmers in these regions generally lack access to uninterrupted cold chains and appropriate cold storage facilities.

To address these challenges, several low-tech alternatives are being explored that can function in areas with limited electricity and are more feasible for small-scale farming operations.

Technology	Description	Temperature provided	Fuel
Passive/ evaporative cooler	works in dry, low humidity areas and does not require electricity	10-25°C	No fuel Architectural measures (shade creation, fountains, etc.)
Absorption refrigerator	Heat-driven technology	below 10°C	Solar Kerosene
Refrigerators	Electricity-driven and therefore Dependent on reliable electricity supply	0°C	Grid Diesel Renewable sources (hydro, solar, batteries, etc.)

Several factors must be taken into account when selecting both the technology and site for a cold storage facility:

- **Location:** The facility should be located in an easily accessible area, preferably close to market hubs and/or produce collection points to ensure convenience and efficiency.
- **Electricity:** It is crucial to ensure reliable access to electricity for the cold storage facility, as this directly influences the facility's size and the type of equipment that can be used. A consistent electricity supply is a key factor in determining the feasibility of the storage operations.
- **Drainage and Waste Disposal:** Proper drainage systems and waste management infrastructure should be in place to manage any waste and prevent contamination, ensuring smooth operations.
- **Water Supply:** A reliable water supply is essential, especially for sanitation purposes, while ensuring that wastewater generated during operations is effectively disposed of.
- **Economic Viability:** The financial returns generated by the cold storage facility should be able to cover the investment and operational costs, making it economically justifiable in the long term.

Overall, selecting the right cold storage technology depends on the required temperatures, reliability, and available resources, such as skilled labor, capital, and a consistent supply of electricity and water.

India's Cold Storage Situation and Prospects

India's estimated annual production of fruits and vegetables is approximately 130 million tonnes, making up about 18% of the total agricultural output. The country's diverse agro-climatic conditions and the increasing availability of effective agricultural practices are driving growth in production. While there is significant potential for further growth in production, the lack of cold storage and cold chain facilities remains a major obstacle in fully harnessing this potential. Currently, the existing cold storage infrastructure is typically designed for single commodities like potatoes, oranges, apples, grapes, pomegranates, and flowers, which leads to inefficient utilization of storage capacity.

Storage of Foods and Storage Conditions

The preservation of food and other perishable commodities through low-temperature storage helps slow down the activities of microorganisms such as bacteria, yeasts, and molds, which are the primary spoilage agents. While low temperatures do not eliminate these microorganisms as high temperatures do, they significantly reduce their activities, providing an effective means of preserving perishable foods in their natural state, something that would be impossible through heating alone. The exact temperature needed for preservation varies depending on the required storage duration, commonly classified as short-term or long-term storage, and the type of product being stored.

In general, there are three categories of products:

1. Living Foods: These include fruits and vegetables, which are alive during storage, distribution, and sale.

2. Processed Foods: These are non-living foods that have been altered in some way, such as meat and fish products.

3. Commodities for Controlled Temperature Storage: These include products like beer, tobacco, and khandsari, which benefit from storage at controlled temperatures.

Living foods, such as fruits and vegetables, naturally resist microbial activity to some extent. The best method for preserving these products is to maintain their vitality while slowing down the natural enzyme processes, which in turn slows ripening or maturation.

Preserving non-living foods is more challenging since they are more vulnerable to spoilage. The focus is on preventing decay and putrefaction of dead tissues. Long-term storage of meat and fish products is typically achieved through freezing, followed by storage at temperatures below -15°C. Only certain fruits and vegetables are suitable for freezing, and it is essential to carefully monitor the recommended storage temperature and humidity. Any deviation from these guidelines can adversely affect the stored products, potentially resulting in total loss.

Some fruits, such as apples, tomatoes, and oranges, cannot be frozen and require precise temperature control for long-term storage. Additionally, some products benefit from storage under controlled or modified atmospheric conditions. Dairy products, which are derived from animal fats, are also considered non-living foodstuffs. They are susceptible to oxidation and fat breakdown, leading to rancidity. Packaging that excludes air and oxygen can help extend the shelf life of these products.

Table 9.3 Desired Storage Environment of Fruits and Vegetables in the Cold Storage

Commodity	Temperature (oC)	Relative Humidity (%)
Apple	-1 - 3	90 - 98
Apricots	-0.5 - 0	90 - 95
Avocado	7 - 13	85 - 90
Asparagus	0 - 2	95 - 97
Beans, green	4 - 7	90 - 95
Beet root	0 - 2	95 - 97
Broccoli	0 - 2	90 - 95
Black berry	-0.5 - 0	95 - 97
Cabbage	0 - 2	90 - 95
Carrots	0 - 2	90 - 95

Commodity	Temperature (oC)	Relative Humidity (%)
Cauliflower	0 - 2	90 - 95
Cherries	0.5 - 0	90 - 95
Cucumber	7 - 10	90 - 95
Brinjal	0 - 2	90 - 95
Grapes	-1 - 1	85 - 90
Lemons	4 - 15	86 - 88
Lettuce	0 - 1	95 - 98
Lime	3 - 10	85 - 90
Mango	11 - 18	85 - 90
Melon water	2 - 4	85 - 90
Orange	0 - 10	85 - 90
Peach	-1 - 1	88 - 92
Potato	1.5 - 4	90 - 94

Utilizing Technology in Cold Storage

A cold storage facility relies on a refrigeration system to maintain the ideal environmental conditions for the products being stored. The refrigeration system operates based on two main principles:

- Vapour Absorption System (VAS)
- Vapour Compression System (VCS)

While the VAS is generally more expensive upfront, it proves to be more economical in the long run due to its lower operational costs, making it a good choice where energy conservation is important. However, VAS has its limitations, particularly when temperatures need to be maintained below 10°C, which is required for many fruits and vegetables (such as mangoes and seeds) for extended storage periods.

On the other hand, VCS is a more cost-effective option compared to VAS. There are three variations of VCS systems, determined by the cooling setup in the storage rooms: diffuser type, bunker type, and fin coil type. The diffuser type is more expensive and is typically used for storage rooms with lower ceiling heights, but it comes with higher operational costs. The bunker type is the least expensive and is favored for rooms with a height exceeding 11.5 meters, offering lower operational costs as well. The fin coil type, which is about 5% more expensive than the bunker type, is highly energy-efficient with low operational costs and provides more space for storing produce. It is most suitable for rooms with heights starting from 5.4 meters.

In refrigeration systems, refrigerants play a crucial role by absorbing heat through evaporation at low temperatures and pressures in the storage area and releasing it through condensation at higher temperatures and pressures in the condenser. Previously, freon was commonly used as a refrigerant; however, due to its environmental impact, its use is being phased out, with a complete ban expected by 2008. As a result, ammonia is increasingly being adopted as a preferred refrigerant, particularly for cold storage units handling horticultural and plantation products.

In a cold storage design,the following heat load factors are typically taken into account:

- Wall, floor and ceiling heat gains due to conduction
- Wall and ceiling heat gains from solar radiation
- Load due to ingression of air by frequent door openings and during fresh air charge.
- Product load from incoming goods
- Heat of respiration from stored product
- Heat from workers working in the room
- Cooler fan load
- Light load
- Aging of equipment
- Miscellaneous loads, if any

10

Cold Chain Logistics and Supply Chain Management System

The continuous flow of a particular thermal profile during the production, packaging, and delivery of a product that is sensitive to temperature is known as the "cold chain." Consider the cold chain as a supply chain with temperature control. A continuous and reliable chain of refrigerated manufacturing, production, packaging, transportation, storage, and distribution is necessary for a cold chain to be successful.

India's cold chain logistics market is expected to grow at an exponential rate due to government initiatives, technological advancements, rising pharmaceutical needs, and consumer demand for fresh products. These factors will ensure efficiency, reduce waste, improve infrastructure, and make significant economic contributions, transforming the country's supply chain landscape.

The shipping environment, package design, shipment duration, and key control points throughout the process must all be understood and managed in order to maintain the cold chain. The technology and procedure that enable the safe movement of products and goods that are sensitive to temperature along the supply chain is known as cold chain logistics. It makes extensive use of science to assess and account for the relationship between perishability and temperature.

Any product that is labeled as "perishable" or known to be so will probably require cold chain management. Foods like meat and seafood, produce, medications, and medical supplies may fall under this category.

LOGISTICS OF THE COLD CHAIN

Cold chain logistics is a vital process that underpins industries reliant on temperature-sensitive transportation, ensuring that products such as fresh produce and essential vaccines are delivered in optimal condition. Grasping the fundamentals of cold chain logistics is crucial to recognizing its role in maintaining product integrity.

Cold chain logistics refers to the highly controlled system of storing, transporting, and handling goods that are sensitive to temperature fluctuations, aiming to preserve their quality and safety. This process involves the use of specialized equipment like refrigerated vehicles and temperature-controlled storage facilities to maintain a consistent environment for the products.

The importance of cold chain logistics is most evident in sectors like food, pharmaceuticals, and chemicals, where the quality of goods such as fresh produce, life-saving vaccines, and volatile chemicals must be preserved until they reach their destination. By preventing temperature variations, cold chain logistics ensures that products retain their integrity, reducing losses and making it an indispensable part of modern supply chains. For time-sensitive and perishable products, optimizing cold chain logistics is essential for smooth operations.

Cold chain logistics consists of several key components, each playing a critical role in the system:

- **Cold Storage:** Facilities designed to hold products that are awaiting transport while maintaining the necessary temperature conditions.
- **Cooling Systems:** These systems are responsible for cooling products to the required temperature and ensuring that they remain at this temperature throughout the supply chain, including during processing, storage, and transportation.
- **Cold Transport:** Ensures that goods are transported under stable temperature and humidity conditions, preventing fluctuations that could compromise quality.
- **Cold Processing:** Specialized facilities that process goods under stringent sanitary conditions to ensure they remain safe and of high quality.
- **Cold Distribution:** Involves the efficient handling and loading of goods, often in boxes or pallets, to ensure they are properly distributed while maintaining the required temperature.

Management of Cold Chain Logistics

Cold chain logistics pertains to the transportation and storage of perishable and temperature-sensitive products within environments that are carefully controlled to maintain specific conditions. These operations are particularly intricate, as products must remain within their designated temperature range at all times. Preventing any temperature deviations is critical to safeguarding the quality and efficacy of the goods. Continuous monitoring and precise temperature control are essential for managing these temperature-sensitive products effectively.

Common Use Cases of Cold Chain Logistics

Cold chain logistics is vital for industries that deal with goods that have a limited shelf life:

- **Food and Beverage Industry:** Globally, the restaurant and grocery sectors depend on cold storage to maintain the freshness of produce, dairy, and frozen foods. Cold chain logistics prevent spoilage, ensuring that consumables retain their nutritional value and farm-to-table quality.

- **Pharmaceutical Products:** A significant volume of medications, including vaccines and biological materials like organs, require specific temperature conditions to remain effective. Cold chain logistics ensure that these products are delivered to healthcare providers under the ideal conditions, which is crucial for saving lives.
- **Chemical and Industrial Products:** Certain chemicals and industrial materials need to be kept at precise temperatures to preserve their stability and effectiveness. Maintaining these specific environmental conditions and constant temperature monitoring is essential during both transportation and storage.

Process of Cold Chain Logistics

Cold chain logistics involves a series of interconnected processes designed to ensure that goods remain within the required temperature range, minimizing the chances of spoilage. The entire process begins with placing goods in facilities equipped with temperature control systems to maintain the appropriate conditions.

Temperature Control

The cornerstone of cold chain logistics is maintaining the right temperature in both storage and transit throughout the supply chain. This involves the use of refrigeration or freezing techniques, depending on the particular needs of each product. Refrigeration typically ranges from 2°C to 8°C, while freezing conditions are maintained below -18°C.

Specialized Equipment

Cold chain logistics depends on specialized vehicles, storage facilities, and shipping containers fitted with refrigeration units and insulation. Refrigerated trucks, cold storage warehouses, and insulated containers are used at various stages of the supply chain to ensure that products stay within the required temperature ranges.

Monitoring and Tracking

Modern cold chain logistics requires constant monitoring of temperature and humidity levels within both storage and transportation units. Companies utilize a variety of Internet of Things (IoT) devices, RFID tags, and real-time GPS tracking systems to collect data on the conditions of the cargo. This allows for immediate corrective actions if any temperature deviations occur, ensuring the products remain intact.

Cold Storage Facilities

After pre-cooling, products are stored in temperature-controlled cold storage units before being transported. These storage facilities maintain specific temperature and humidity levels to preserve the quality of perishable goods.

In India, cold storage units are widely used for fruits, vegetables, dairy, and pharmaceuticals, but there is still a gap in infrastructure, particularly in rural areas.

Distribution and Last-Mile Delivery

Once products reach their destination, they are stored in regional distribution centers or retail cold rooms before being delivered to end consumers. Efficient last-mile delivery using refrigerated vans or insulated packaging is crucial to maintaining product quality until it reaches customers. In urban areas, e-commerce companies and supermarkets rely on cold chain solutions for delivering perishable food and medicines.

Monitoring and Quality Control

Throughout the cold chain process, temperature and humidity levels must be continuously monitored to prevent spoilage. Modern technologies such as GPS tracking, real-time temperature sensors, and data loggers help maintain supply chain transparency and ensure compliance with safety standards.

Packaging Solutions

Packaging plays a crucial role in cold chain logistics, ensuring that products stay at the proper temperature. Notable packaging solutions include insulated containers, gel packs, dry ice, and phase change materials that help maintain the required temperatures over extended periods. Engineers have spent decades perfecting these thermal protection systems to ensure that products arrive in optimal condition.

Technological Innovations

Recent advancements in technology have revolutionized cold chain logistics. Software solutions that integrate with existing transportation management systems (TMS), warehouse management systems (WMS), real-time visibility tools, and AI-driven predictive maintenance have drastically improved how businesses operate. As a result, companies have experienced better risk management and a more streamlined overall supply chain process.

TECHNOLOGIES FOR THE COLD CHAIN

Cold chain transportation employs a variety of methods to maintain the appropriate temperatures for goods during transit. Factors such as the duration of transport, the size of the shipment, and seasonal conditions all influence the method selected. Common methods include:

- **Gel Packs:** Typically used for shipping medical and pharmaceutical products, gel packs are effective at maintaining temperature stability.
- **Dry Ice:** Ideal for keeping goods frozen over long durations, dry ice is commonly used for pharmaceuticals, dangerous goods, and food products. Unlike traditional ice, dry ice sublimates, turning directly from solid to gas without melting.

- **Liquid Nitrogen:** Primarily used for transporting biological materials such as organs and tissues, liquid nitrogen offers extremely low temperatures that keep items frozen for extended periods.
- **Eutectic Plates:** Also known as cold plates, these are similar to gel packs but can be reused, making them a more sustainable option for temperature control during transport.
- **Reefers:** These are temperature-controlled, insulated containers (including vans, trucks, or ISO containers) that circulate temperature-controlled air, ensuring the proper environment is maintained throughout transit.
- **Quilts:** Insulated quilts are placed around or over freight to help maintain consistent temperatures, ensuring that frozen items stay frozen for longer periods.

The cold chain adheres to a meticulously planned procedure to guarantee the quality of products that are sensitive to temperature changes. Here is a summary of how it works to keep everything at ideal levels. The cold chain functions as a meticulously planned system intended to preserve the integrity of a product. It includes:

- **Temperature Monitoring:** Real-time condition tracking is ensured by sophisticated sensors and Internet of Things-enabled devices.
- **Specialized Transportation:** Goods that are sensitive to temperature can be handled by refrigerated trucks, ships, and air cargo carriers.
- **Storage Options:** To keep goods safe while in transit, warehouses and distribution hubs use freezers and cold rooms.

The cold chain reduces spoilage by combining these components, guaranteeing that goods stay safe and functional. In a similar vein, supply chain logistics optimization can assist companies in better managing transportation and inventory while guaranteeing temperature control throughout the supply chain.

Important Elements of Logistics in the Cold Chain

Every component of the cold chain is essential to preserving temperature regulation. Below are the main elements:

- **Cold Chain Packaging:** Keeping products at the proper temperature requires efficient packaging. Common solutions for providing thermal protection during transit include dry ice, gel packs, and insulated containers.
- **Cold Chain Storage:** It is crucial to have warehousing facilities with temperature-controlled spaces. These areas consist of freezers, cold rooms, and humidity-controlled spaces designed to meet the requirements of particular products.
- **Tracking and Shipping in the Cold Chain:** GPS tracking in conjunction with refrigerated transportation guarantees that goods stay in the proper condition during the trip. If deviations happen, real-time monitoring enables prompt intervention.

Cold chain logistics success is largely dependent on distribution and transportation management, as is the case with all facets of logistics. These crucial tasks are streamlined by effective systems.

COLD CHAINS STANDARDS & REGULATIONS IN INDIA

For perishable goods like fruits, vegetables, dairy products, meat, seafood, and medications to remain safe and of high quality, cold chain logistics is essential. Reducing post-harvest losses and guaranteeing food security in India require an effective cold chain. The Indian government and several regulatory agencies have put in place rules and regulations that address distribution, storage, and transportation in order to control and standardize cold chain operations.

Regulatory Bodies Governing Cold Chain in India

In India, a number of organizations regulate cold chain operations to guarantee adherence to safety and quality requirements:

- **Food Safety and Standards Authority of India (FSSAI):** Establishes standards for transportation and cold storage and regulates the safety of perishable food items.
- **Bureau of Indian Standards (BIS):** Creates technical specifications for equipment, cold storage facilities, and refrigerated cars.
- **National Centre for Cold Chain Development (NCCD):** Offers policy recommendations and promotes cold chain infrastructure.
- **Agricultural and Processed Food Products Export Development Authority (APEDA):** Oversees the export of agricultural and processed food products, ensuring adherence to cold chain requirements.
- **Central Drugs Standard Control Organization (CDSCO):** Oversees cold chain logistics for pharmaceuticals, especially vaccines and medications that are sensitive to temperature.

Key Standards and Guidelines for Cold Chain Logistics

Temperature Standards for Different Products

Maintaining optimal temperature conditions is crucial for product safety and longevity. India follows globally recognized temperature standards for various perishable items:

- Fresh fruits and vegetables: 0°C to 12°C
- Dairy products: 0°C to 4°C
- Frozen meat and seafood: -18°C or lower
- Vaccines and pharmaceuticals: 2°C to 8°C (for general vaccines), -20°C or lower (for specific drugs)

FSSAI and APEDA have issued guidelines specifying these temperature ranges to ensure food and drug safety.

Cold Storage and Warehousing Regulations

To ensure efficiency and safety, cold storage facilities must follow specific design and operational guidelines. BIS offers guidelines for energy efficiency, insulation, and cold storage construction. Important rules consist of:

- Proper temperature monitoring and control systems
- Adequate ventilation and humidity control
- Backup power supply to prevent temperature fluctuations
- Compliance with hygiene and sanitation norms to prevent contamination

Refrigerated Transport Standards

When transporting perishable goods, refrigerated vehicles, also known as refrigerated trucks, must adhere to BIS and FSSAI regulations. Among the regulations are:

- Use of temperature-controlled units with proper insulation
- Real-time temperature monitoring and data logging
- Regular maintenance and calibration of refrigeration equipment
- Proper loading and unloading practices to prevent temperature fluctuations

Pharmaceutical Cold Chain Regulations

Strict cold chain guidelines are adhered to by the pharmaceutical industry, especially for vaccines, biologics, and insulin. In collaboration with the Ministry of Health, CDSCO requires:

- Use of validated cold storage and transport systems
- Temperature-sensitive labeling and packaging
- Continuous temperature monitoring using data loggers and tracking devices
- Compliance with Good Distribution Practices (GDP) to prevent temperature excursions

Government Initiatives to Strengthen Cold Chain Standards

To strengthen cold chain infrastructure and ensure regulatory compliance, the Indian government has started a number of initiatives:

- **Pradhan Mantri Kisan Sampada Yojana (PMKSY):** Supports the development of cold storage and food processing facilities.
- **Cold Chain Development under the Ministry of Food Processing Industries (MoFPI):** Provides financial assistance for setting up integrated cold chain projects.
- **Faster Adoption of Cold Chain Technology:** Encourages the use of modern refrigeration techniques and IoT-based monitoring systems.

In India, cold chain standards and laws are essential for guaranteeing the dependability, effectiveness, and safety of products that are sensitive to temperature. Despite the existence of regulatory frameworks, problems like poor infrastructure, exorbitant expenses, and difficulties with compliance still exist. Ensuring sustainable growth in this industry will require stepping up enforcement, implementing cutting-edge technologies, and investing more in cold chain logistics.

Common Cold Chain Management Issues

Several issues in cold supply chain management can significantly affect freight shipments. Drivers are typically aware of these potential challenges and strive to address them proactively to avoid problems.

- **Product quality concerns:** For food items and produce, maintaining quality starts before shipping. Proper sanitizing, cleaning, and sorting must be done before packaging and loading.
- **Inadequate packaging:** Effective packaging is critical to prevent contamination and damage during transportation. Proper airflow is also essential to maintain the integrity of the product.
- **Lack of proper documentation:** It is crucial to document every step of the cold supply chain, especially during transit. Data loggers that record temperature and environmental conditions can help ensure that any deviations do not spoil the goods.
- **Shipment/transport delays:** Delays are problematic for any kind of shipment, but they can be particularly harmful in cold chain logistics, where time-sensitive technology is critical.
- **Disrupted climate control or temperatures:** Temperature fluctuations can significantly affect the quality of goods. These fluctuations may occur due to frequent deliveries, improper loading, extreme weather, faulty cooling equipment, or inadequate pre-cooling.

Reasons Why Cold Chain Management is Essential

Products and Goods Integrity

The safety, quality, usability, and integrity of goods and products rely heavily on cold chain management. Modern cold chain practices, utilizing advanced thermal packaging, temperature monitoring, and storage technologies, ensure the safety and integrity of temperature-sensitive products. Whether it's preserving the freshness of milk, keeping ice cream frozen, or protecting vaccines, cold chain management plays a vital role in maintaining the supply chain of essential goods. Customers expect products to arrive without any defects such as bruising, discoloration, textural changes, or bacterial growth, and cold chain management ensures these expectations are met.

Product Safety

Cold chain management helps you avoid negative publicity and costly safety issues. Nobody wants their name associated with unsafe products. Effective temperature control and monitoring systems ensure compliance with both internal standards and regulatory requirements, safeguarding consumers and ensuring product safety. Continuous monitoring of factors like temperature, humidity, air quality, GPS location, energy consumption, delivery, and storage conditions are all integral aspects of contemporary cold chain management.

Cutting Down on Waste and Loss

One of the most significant challenges in any supply chain is preventing waste, damage, and product loss. With cold chain management, products are tracked and monitored throughout each stage of the supply chain. Companies rely on advanced thermal packaging, specialized refrigerants, reefers, cold chain parcel systems, temperature monitoring, and insulated containers to protect their goods. Given the tight margins and external pressures on the supply chain, businesses cannot afford the financial or reputational consequences that result from the loss of temperature-sensitive products.

Improving Efficiency

Transporting goods across the country or globally is a complex, expensive, and time-consuming process. Cold chain management uses integrated, smart IoT-enabled tracking and monitoring systems, allowing businesses to streamline transportation, delivery, and storage. This improved process helps increase operational efficiency while reducing overall costs.

Customer Service

Both B2B and B2C customers expect timely delivery and high-quality, safe products. Cold chain management provides the necessary oversight and control to meet these expectations by ensuring efficient distribution, storage, and delivery. Services like Just in Time (JIT) delivery of thermal packaging supplies and gap analyses for cold chain best practices contribute to a more efficient and cost-effective cold supply chain, benefiting both businesses and customers.

Compliance and Regulatory Standards in India

Preserving perishable items like food, dairy, seafood, medications, and vaccines requires effective cold chain management. Strict adherence to legal requirements guarantees that temperature-sensitive goods in India keep their quality and safety during distribution, storage, and transit. Cold chain operations are regulated by a number of government organizations to avoid contamination and spoiling.

Compliance Requirements

1. **Temperature Control Standards:**
   - **Perishable Food:** From 0°C to 12°C
   - **Frozen Products:** Below -18°C
   - **Dairy and Pharmaceuticals:** From 2°C to 8°C
   - **Vaccines and Biologics:** Strictly maintained at required temperatures using validated storage and transport systems.
2. **Cold Storage and Warehousing:**
   - Compliance with hygiene, ventilation, and energy efficiency standards.
   - Proper insulation and backup power for temperature consistency.

3. **Refrigerated Transport Regulations:**
   - Vehicles must have real-time temperature monitoring and controlled environments.
   - Proper loading, unloading, and handling to prevent fluctuations.
4. Good Distribution Practices:
   - Mandatory for pharmaceutical cold chain logistics.
   - Requires documentation, tracking, and compliance with global best practices.

India's cold chain sector meets international regulatory requirements, minimizes losses, and guarantees product safety by following these guidelines.

Cold Chain Management Process

The cold chain management process encompasses all stages necessary to ensure the safe and secure storage, packaging, tracking, transportation, customs clearance, product management, and delivery of temperature-sensitive products. This process heavily relies on advanced technology to monitor and control temperature, logistics, movement, delivery, and security of goods throughout their journey.

- **Storage:** Effective cold storage is fundamental to cold chain management, involving specialized equipment like insulated containers, vaccine and medical refrigerators, freezers, refrigerants, and cold rooms or chillers. Cold chain storage facilities and equipment are the starting point for maintaining the right conditions for perishable goods.
- **Packaging:** When goods leave storage facilities, the use of cold packaging is crucial to ensure they maintain their temperature and remain secure. Packaging products like cold chain parcels, pallet systems, refrigerant bricks, gel packs, gel bottles, and insulated containers and envelopes are essential for keeping products cold, improving transport efficiency, and minimizing damage, spoilage, and loss.
- **Tracking:** The use of data loggers and temperature indicators allows companies, shippers, and customers to monitor and track temperature and GPS location in real time. This tracking system is vital for maintaining the cold chain and ensuring the integrity and quality of the products.
- **Transportation:** With the global supply chain in mind, temperature-sensitive goods are transported in various ways across the world. Specialized vehicles, such as refrigerated trucks and reefer containers, help maintain the required conditions while goods are moved across air, sea, and land routes.
- **Customs Clearance:** Customs clearance is a critical part of the cold chain, involving necessary paperwork that ensures products can be transported and delivered efficiently. Non-compliance with customs regulations can lead to delays, affecting product quality, causing potential losses, and incurring additional storage, transport, and delivery costs.

- **Product Management:** Handling supplies like cold chain parcels, insulated envelopes, and containers, along with warehouse management practices such as using autonomous mobile robots and forklifts, ensures the secure and safe movement of products within the cold chain.
- **Delivery:** The final step in cold chain management is delivering temperature-sensitive products to B2B and B2C customers. This involves a well-coordinated and efficient delivery process, which includes reviewing documentation, following best practices for internal handling, managing risks, and securing compliant storage, warehouse, and display systems.

FUTURE TRENDS IN COLD CHAIN LOGISTICS

Cold chain logistics is undergoing rapid transformation as it adapts to new technologies and evolving market demands. The following trends are expected to define the future of cold chain logistics:

Green Technologies and Sustainability Initiatives

As sustainability becomes a top priority for businesses and consumers alike, cold chain logistics is embracing eco-friendly practices. The adoption of renewable energy sources, such as solar power for refrigerated storage facilities and electric vehicles for transportation, helps reduce carbon footprints.

Additionally, the use of eco-friendly packaging materials and energy-efficient cooling systems is gaining traction. Businesses are increasingly prioritizing sustainability by reducing energy consumption and emissions throughout the cold chain. This shift towards greener logistics also aligns with consumer preferences for environmentally conscious brands and supports regulatory compliance for businesses in regions with stringent environmental standards.

Modern Automation for Cold Chain Operations

The rise of automation is revolutionizing cold chain logistics. Autonomous vehicles, including drones and self-driving trucks, are being deployed for last-mile delivery. These innovations allow for faster, more efficient delivery of temperature-sensitive goods, especially in areas with high demand.

Automated warehouses are becoming commonplace, where robots handle sorting, packaging, and even temperature monitoring. These systems not only improve speed and accuracy but also reduce human error and labor costs.

With predictive analytics and AI integration, logistics companies can better forecast demand, optimize delivery routes, and adjust temperature controls in real-time to ensure product quality is maintained throughout the supply chain.

Real-time Monitoring and Greater Digitization

The future of cold chain logistics is heavily reliant on digital transformation. Technologies like AI, IoT, and blockchain are enabling end-to-end visibility and control over the entire cold chain. IoT sensors, for example, continuously monitor temperature and humidity levels during transit, providing real-time data that allows for quick action if conditions deviate from set parameters.

Blockchain enhances transparency by offering an immutable record of every transaction and movement in the supply chain, ensuring greater accountability and security. Meanwhile, AI-powered systems help with demand forecasting, route optimization, and anomaly detection, allowing businesses to preemptively address potential issues before they affect the integrity of goods.

Risk Management and Resilience in the Supply Chain

As cold chain logistics grows more intricate, businesses must be ready to handle a broader array of risks. Global supply chains are becoming more susceptible to disruptions, ranging from natural disasters to geopolitical conflicts.

To maintain the stability of cold chain systems, businesses will need to adopt effective risk management strategies, such as real-time tracking and predictive analytics. These tools will be essential in identifying and addressing potential risks, helping companies reduce disruptions and ensure consistent service levels.

Moreover, integrating business continuity planning into the cold chain strategy will be crucial in minimizing the impact of risks that could threaten the delivery of temperature-sensitive goods.

For businesses aiming to stay competitive, exploring advanced supply chain risk management approaches is vital, particularly as new and complex risks arise in today's interconnected logistics landscape.

Cold Chain vs. Supply Chain

The below table differentitates between cold chain and supply chain:

Aspect	Cold Chain	Supply Chain
Definition	A supply chain that specifically manages temperature-sensitive goods.	A network of processes to produce and distribute goods from origin to consumption.
Products	Perishable items such as food, pharmaceuticals, and biological products.	A wide variety of goods including raw materials, manufactured products, and consumables.
Temperature Control	Critical – requires temperature-controlled storage and transportation.	Not necessarily temperature-sensitive, though some products may need temperature control.
Technology Used	Refrigerated trucks, cold storage, temperature monitoring devices, IoT sensors.	Standard logistics technology like inventory management, transportation management, and tracking systems.

Aspect	Cold Chain	Supply Chain
Purpose	To maintain the quality, safety, and integrity of temperature-sensitive products.	To ensure timely, cost-effective delivery of goods from producers to consumers.
Key Processes	Storage, packaging, temperature monitoring, transportation, and delivery.	Sourcing, manufacturing, storage, transportation, and distribution.
Risk Management	Focuses on maintaining a constant temperature to prevent spoilage or degradation.	Focuses on reducing costs, optimizing efficiency, and managing delays and risks.
Compliance	Strict regulations for temperature monitoring, packaging, and transportation.	Regulatory requirements depend on the type of goods, but not typically related to temperature.
Disruptions	Disruptions like temperature fluctuation can spoil products and cause significant losses.	Disruptions like supply delays or shortages can impact the flow of goods but may not result in product loss.
Example Industries	Food and beverages, pharmaceuticals, medical products, floral products.	Retail, manufacturing, automotive, electronics, consumer goods.

Temperature-Sensitive Products Need Cold Chain Management

Any product that is temperature sensitive needs cold chain management. These temperature-sensitive products need cold chain management as follows:

- Vaccines including the COVID-19 vaccines from Pfizer and Moderna.
- Medical products including blood, biologics, insulin, and other medical devices.
- Wine requires consistent cool temperatures to prevent premature aging and loss of flavor.
- Frozen desserts including ice cream and gelato need to be kept frozen to prevent bacterial growth, crystallization, and loss of flavor.
- Perishable meat, seafood, and poultry is highly susceptible to bacterial growth when it is not kept below 40° F or frozen at 0° F.
- Fresh and frozen fruits and vegetables are at risk of skin damage, bacterial growth, flavor loss, bruising, and more when the cold chain is not maintained.
- Milk, milk alternatives, and dairy products including yogurt, oat milk, soy products, and fermented dairy all require cold chain management. As an example, milk must be processed, packaged, transported, and stored at below 45° F to prevent bacterial growth.

Cold Chain Management Challenges

Running a cold chain involves overcoming several key difficulties. Let's look at some of the most significant challenges businesses encounter.

- **Managing Temperature Variations:** Keeping temperatures consistent can be difficult due to equipment malfunctions, human mistakes, or external environmental factors. Advanced technologies such as IoT sensors and predictive analytics can help minimize these challenges.
- **Adhering to Compliance and Regulations:** Cold chain logistics must meet strict industry regulations. For example, the pharmaceutical sector follows standards like Good Distribution Practice (GDP) to guarantee the safety and integrity of products.
- **Managing Costs:** Operating a cold chain comes with high costs, such as the need for specialized equipment, energy consumption, and skilled labor. Reducing costs involves optimizing operations and adopting more energy-efficient solutions. Finding ways to reduce logistics costs can help in controlling the overall expenses tied to cold chain management.

Best Practices in Cold Chain Logistics Management

To enhance operational efficiency and reduce risks, businesses can implement best practices. This section highlights essential strategies to improve cold chain performance:

- **Invest in Training:** Ensure that staff are well-trained in handling temperature-sensitive products to avoid errors and maintain quality.
- **Embrace Technology:** Implement automated systems that provide real-time monitoring and data reporting, enhancing operational transparency.
- **Optimize Transportation Routes:** Strategically plan delivery routes to shorten transit times and minimize potential risks, ensuring goods arrive on time and in good condition.
- **Collaborate with Experts:** Work with experienced cold chain service providers to streamline operations and improve logistical effectiveness.

An important component of cold chain management is also improving inbound transportation and understanding how supply chain tracking enhances logistics efficiency. This process significantly boosts visibility and control, ensuring the cold chain operates smoothly from start to finish.

SUPPLY CHAIN MANAGEMENT SYSTEM

A "Post-harvesting Supply Chain Management System" is a holistic system that manages the handling, storage, processing, packaging, and transportation of agricultural products from harvest to consumption. Its main objective is to preserve quality, reduce losses, and guarantee food safety across the entire supply chain.

Managing the Supply Chain for Fresh Produce

Post-harvest supply chain management is especially challenging in the fresh produce industry, where mistakes are far less forgiving than in many other sectors.

A misplaced shipment in a warehouse for a week isn't just an inconvenience – it can result in complete loss of the entire shipment. Managing the fruit supply chain adds an even greater challenge, particularly in fruit storage environments. The accumulation of ethylene, a natural agent that causes fruit to ripen, can be a significant issue. Without proper control, ethylene buildup can lead to accelerated spoilage and waste.

However, solutions exist to address this problem. For instance, Post Harvest's advanced ripening sensor collects ethylene samples from the air, providing early warnings when action is necessary and improving the accuracy of ripening forecasts.

Effective supply chain management hinges on identifying opportunities for continuous improvement. When these improvements accumulate, their combined effect can be transformative. In any business, efficient supply chain management directly impacts sales, revenues, profits, customer satisfaction, and more. In the fresh produce industry, failing to manage the supply chain effectively can have serious consequences.

Stages of Food Supply Chain

The food supply chain consists of four distinct stages, each vital to ensuring that food reaches consumers safely and efficiently. These stages are defined based on factors such as whether the food requires processing, packaging, and transportation, as well as its consumption level. To understand how the food supply chain operates, it's helpful to break down each stage, starting with production.

Production

The food supply chain begins at the production stage, where food is sourced. This can involve farming, growing, or raising food in compliance with both local and international regulations. These rules ensure that food products meet the required standards for quality, appearance, and safety before they move forward in the supply chain.

Handling and Storage

After food products are harvested, they undergo various handling procedures to prepare them for the next phase. For example, some foods like potatoes need to be washed before packaging, while others may be directly packed and shipped. In some cases, the products are sent to processing plants to be turned into different forms, such as chips or crisps. The handling methods vary depending on the food and its intended destination.

Processing and Packaging

Before food products can proceed to packaging, they must meet strict safety standards. Once they have passed the required safety checks, they are packaged appropriately to protect their quality and prepare them for further distribution in the supply chain.

Distribution

Once packaged, the food is shipped to its final destinations, which often include markets, grocery stores, or other parts of the food industry. Depending on the product, food can be transported by different modes of transport, including ships or airplanes. The distance from producer to consumer is referred to as "food miles," which measures the environmental impact or "food print" associated with producing and transporting the food.

Different Types of Food Supply Chain Models

Below are the distinct models used in food supply chains, each selected by businesses based on their unique needs and operational goals.

Continuous Flow Model

This model is commonly applied in high-demand production environments because it ensures stability in inventory and raw materials. It is particularly well-suited for established brands with mature supply chains where the customer base remains relatively constant and predictable. The continuous flow model helps maintain a steady, uninterrupted supply of products to meet consistent demand.

Fast Chain Model

The fast chain model is designed for businesses dealing with trendy, perishable products. It enables companies to quickly adapt to changing consumer preferences by sourcing fresh ingredients and developing new dishes in response to emerging food trends. Similar to the fashion industry, food trends can change rapidly, and restaurants and foodservice providers use fast supply chains to introduce new products and stay ahead of the competition. Once a trend fades, the supply chain is adjusted to focus on the next popular food trend.

Efficient Chain Model

This model is particularly useful for companies operating in highly competitive industries where maintaining efficiency is crucial. The efficient chain model emphasizes minimizing costs while maximizing productivity in both logistics and production processes. By focusing on streamlining operations and reducing unnecessary expenses, businesses can improve their profit margins. It's particularly beneficial in sectors where competition is fierce, and profit margins are thin, allowing companies to optimize resources and maintain a steady supply for vendors and customers.

Agile Model

The agile model in food supply chains involves four main components: virtual integration, process alignment, network base, and market sensitivity. Virtual integration allows businesses to track real-time changes in market demand, while process alignment ensures that all supply chain operations are synchronized and that responsibilities are shared across the business. The network base component ensures equal involvement from all participants in the food supply chain. Market sensitivity allows businesses to quickly adjust production rates based on fluctuations in food demand, ensuring responsiveness to market shifts.

Custom-Configured Model

This model merges the agile and continuous flow models to allow for greater flexibility in handling diverse product configurations in the food supply chain. It is particularly suitable for situations where businesses offer customizable products or services. This hybrid model helps food businesses remain adaptable to changing customer preferences and ensures they can meet varied consumer needs effectively.

Flexible Model

The flexible model is designed to help food businesses manage varying levels of demand, including both peak and low-demand periods. This model consists of three main components: part segmentation, accurate stocking algorithms, and flexible planning. By using automation and diversifying suppliers, businesses can adjust to changes in demand while optimizing inventory management and production efficiency. This ensures that food businesses can maintain optimal stock levels and production capabilities, even when demand fluctuates significantly.

Challenges in Food Supply Chain

The food supply chain faces numerous vulnerabilities that can lead to significant disruptions with far-reaching consequences. These disruptions can cause food shortages, price increases, and a range of other issues. Below are some key factors that can disrupt the food supply chain:

Health Crisis

Health crises, such as the COVID-19 pandemic, can severely affect the workforce, impacting critical areas like farming, manufacturing plants, and grocery stores. A shortage of workers can cause major disruptions to the food supply chain's efficiency. Additionally, health outbreaks or contamination events at particular farms can trigger product recalls and result in waste, reducing the availability of safe food for consumers.

Extreme Weather Conditions

Severe weather events, including storms, floods, and blizzards, can cause widespread damage to factories, farms, and transportation infrastructure. This damage disrupts the ability to produce and distribute food, leading to delays and shortages in the supply chain. Such extreme weather conditions often prevent goods from reaching consumers in a timely manner, exacerbating the challenges faced by the food industry.

Energy Security

Energy is essential at every level of the food supply chain. Farmers rely on energy to cultivate crops and maintain livestock, while manufacturers depend on electricity for processing and packaging food products. Additionally, transportation services require fuel to deliver goods. Disruptions in the energy supply, whether through shortages or high costs, can have a profound effect on the food supply chain, leading to scarcity, delays, and rising consumer prices.

Geopolitical Conflict

Geopolitical conflicts present significant risks to the stability of the food supply chain. When a country involved in a conflict is a major producer of certain food products, damage to infrastructure and workforce shortages can disrupt production. These disruptions can halt both domestic and international food trade, causing a ripple effect throughout the global food supply network.

BEST PRACTICES FOR FOOD SUPPLY CHAIN MANAGEMENT

Supply Chain Visibility

Visibility is crucial in today's consumer-driven market. Modern consumers are increasingly concerned about where their food comes from, even down to the smallest ingredients. Data accuracy plays a key role in providing this visibility, while also enhancing food safety and maintaining brand integrity. Without adequate visibility, blind spots and unnecessary risks can develop, undermining consumer trust, which in turn can lead to reduced profitability and increased legal liabilities. One of the main reasons for the lack of visibility in the food supply chain is the absence of robust digital infrastructure. This is where blockchain technology comes into play, offering a promising solution to enhance supply chain visibility and traceability. Blockchain allows for real-time visibility of transactions, and once information is recorded on the blockchain, it becomes permanent and tamper-proof. This technology provides producers, suppliers, distributors, retailers, and consumers with reliable, accessible information, such as the origin of raw materials, supplier details, product conditions, and more.

Maintaining the Safety and Quality

Ensuring product safety and quality is becoming increasingly challenging for manufacturers today. Issues such as improper storage, transportation delays, and adverse weather conditions can significantly impact the integrity of food products.

The key to addressing these challenges lies in careful selection of raw materials, adherence to global production standards, and comprehensive testing procedures. Collaborating with accredited laboratories equipped with advanced technology ensures that quality assurance is maintained at the highest level. Improved packaging methods also play a critical role in preserving both product freshness and safety. Furthermore, partnering with experienced logistics providers, especially those specializing in refrigerated goods, allows for real-time tracking and temperature regulation, further safeguarding product quality throughout the supply chain.

Collaboration

Inadequate collaboration significantly disrupts the food supply chain, leading to errors, waste, and a lack of trust among stakeholders. This problem is particularly pronounced in global operations. To address this, advanced supply chain collaboration software has been developed to streamline cooperation among all supply chain partners. By utilizing state-of-the-art technology, this software allows organizations to collaborate in real time with material suppliers, freight forwarders, logistics service providers, and customs authorities.

With the integration of advanced AI, analytics, and machine learning, the software facilitates quick, action-oriented solutions, boosting collaboration and ultimately improving profitability.

Inventory Tracking

Effective inventory management is crucial for controlling costs, maintaining product quality, and ensuring customer satisfaction. The key lies in finding the perfect balance: excessive inventory leads to spoilage and waste, while insufficient inventory results in customer disappointment. Achieving customer satisfaction while keeping inventory levels low remains a continuous challenge. The solution can be found in modern inventory management systems. These systems provide real-time visibility into inventory at every stage of the supply chain, whether it's on-site, off-site, or in transit. By utilizing advanced technologies such as RFID, IoT, telematics, and automated tracking, these tools ensure precise and up-to-date inventory data.

Furthermore, integrating the supply chain with sales data from stores or restaurants allows for greater control and helps reduce excess inventory. Advanced network solutions can detect shifts in demand at the point of sale, automatically adjusting inventory levels and generating orders in real-time. This optimization ensures that inventory levels are just right, service levels remain high, customers are satisfied, and waste is minimized.

11

Quality Management and Tracking Food Supply Chain

A Quality Management System (QMS) is an organized framework that outlines and documents the processes, procedures, and responsibilities necessary for an organization to achieve its quality policies, objectives, and practices. The primary aim of a QMS is to minimize waste, enhance efficiency, and boost customer satisfaction. At its foundation, Total Quality Management (TQM) is a business philosophy that stresses the importance of customer satisfaction and loyalty as the key drivers for long-term business success. TQM involves the collaboration of all stakeholders within the company to enhance processes, products, services, and the overall culture of the organization.

While TQM may appear to be a natural approach, it emerged as a groundbreaking concept. In the 1920s, businesses began to increasingly rely on statistical theory, and in 1924, the first-known control chart was created. As people built on the concepts of statistics, they eventually developed the method known as statistical process control (SPC). However, it wasn't until the 1950s that SPC was successfully implemented in business practices. Quality management and tracking in the food supply chain are essential for ensuring food safety, reducing wastage, and maintaining regulatory compliance. It involves monitoring every stage, from production and processing to storage, transportation, and distribution.

Key quality management practices include temperature control, hygiene standards, and proper packaging to prevent contamination. Cold chain logistics play a crucial role in preserving perishable goods by maintaining optimal storage conditions. Regulatory bodies like FSSAI and APEDA enforce guidelines for food safety, ensuring adherence to quality standards. Tracking technologies such as RFID, IoT sensors, and blockchain enhance transparency and traceability. These systems provide real-time data on temperature, humidity, and product movement, allowing stakeholders to detect and prevent potential issues. End-to-end tracking ensures accountability and quick response to recalls or quality concerns, thereby strengthening consumer trust and improving the overall efficiency of the food supply chain.

Figure 11.1 Quality management principles

QUALITY MANAGEMENT SYSTEM

A quality management system (QMS) is a structured framework that outlines and records an organization's processes, procedures, and responsibilities for achieving quality standards, practices, and goals. It enables the organization to manage and align its activities to fulfill customer and regulatory requirements while continuously enhancing its overall effectiveness and efficiency.

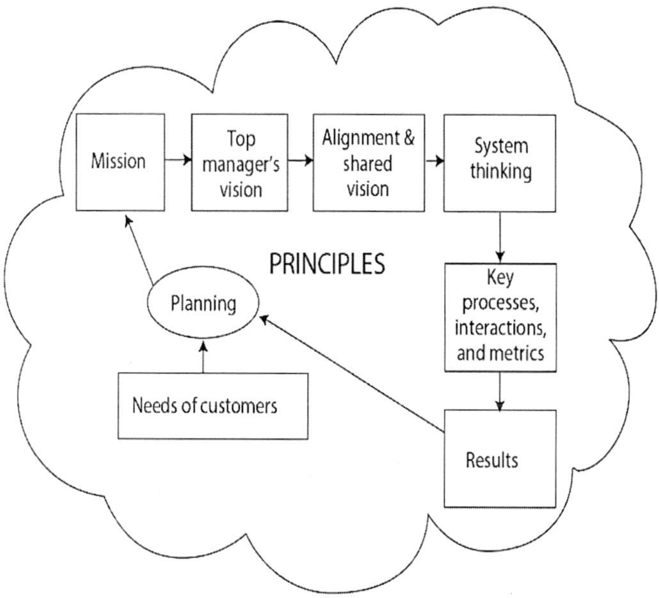

Figure 11.2 Quality management system

Benefits of a QMS

Implementing a QMS influences every aspect of an organization's performance. The advantages of having a documented QMS include:

- Fulfilling customer requirements, which builds trust in the organization, leading to increased customers, sales, and repeat business.
- Meeting organizational requirements, ensuring compliance with relevant regulations, and delivering products and services in the most cost-effective and resource-efficient way, paving the way for growth, expansion, and profitability.

These benefits also bring additional advantages such as:

- Defining, refining, and controlling processes.
- Reducing waste.
- Preventing errors.
- Lowering costs.
- Identifying and supporting training opportunities.
- Engaging employees.
- Setting a clear organizational direction.
- Demonstrating a commitment to achieving consistent results.

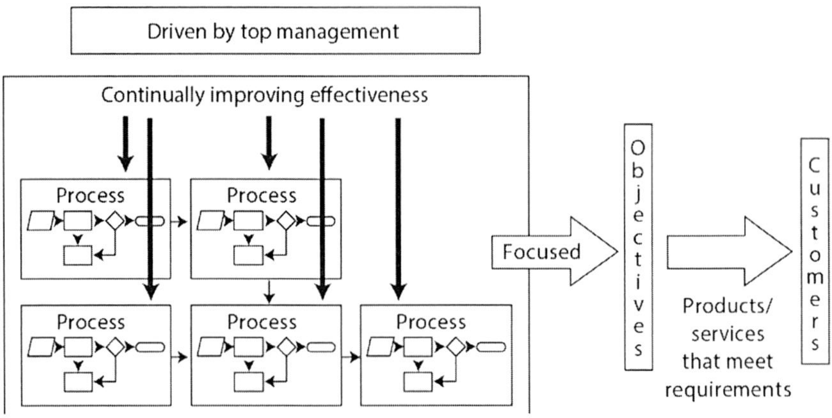

Figure 11.3 Network of Interrelated Process

Effective Crop and Animal Production

Implementing quality assurance measures begins at the farm level with the adoption of good agricultural practices (GAPs) and good veterinary practices (GVPs). GAPs are practices that focus on producing food that is safe, high-quality, environmentally sustainable, and properly handled, stored, shipped, and managed throughout the production process. When GAPs are applied correctly during the cultivation of primary food crops, consumers can be confident that the food will meet safety and quality standards at the time of harvest. Some examples of GAPs include:

- Selecting the most suitable land for food crop production.
- Planting high-quality seeds of appropriate varieties.
- Using authorized and acceptable chemical inputs (fertilizers, pesticides) in accordance with approved guidelines (such as concentration, frequency, and timing of application).
- Ensuring irrigation water quality (if used).
- Employing proper harvesting, on-farm storage, and handling techniques.
- Implementing appropriate methods for shipping to markets or food processors.

Similarly, GVPs are designed to guarantee that food derived from animals meets acceptable quality and safety levels. These practices serve as the foundation for professional veterinary care and treatment of animals, particularly those intended for human consumption. Key GVP-related measures include:

- Ensuring only healthy animals are slaughtered for human food.
- Using drugs to control animal diseases only when they are safe for their intended use and according to approved guidelines (i.e., correct dosages, frequency, and timing), and ensuring no unsafe drug residues remain in edible tissues at the time of consumption.
- Ensuring chemicals used in animal husbandry (such as insect pest control treatments) are safe and applied in accordance with instructions, with no unsafe residues left in edible tissues.
- Properly conducting live animal inspection and handling before slaughter and inspecting and handling carcasses post-slaughter.
- Maintaining appropriate temperature controls, storage conditions, handling, and sanitary practices during processing and butchering to prevent contamination.
- Ensuring shipping and handling practices prevent unnecessary exposure to contamination.

When implemented correctly, both GAPs and GVPs serve to protect food at the primary production stage from contamination caused by foreign materials (such as filth, decomposed substances, rocks, dirt, and sand); toxic chemicals and environmental contaminants (including heavy metals, pollution, and industrial chemicals); excessive or unsafe agricultural chemical residues (such as pesticides, fertilizers, and veterinary drugs); damage or contamination by pests and vermin; and biological contaminants (such as mold, bacteria, or viruses). These factors can lead to spoilage, crop damage, and foodborne illnesses or chronic health issues in humans.

Additionally, consuming animal products from animals fed contaminated feed may lead to human health risks if those contaminants carry over into the edible meat.

Good Practices in Processing

Ensuring the quality and safety of food designated for manufacturing or processing involves the application of good manufacturing practices (GMPs) and good hygienic practices (GHPs). When these practices are properly implemented, they guarantee safety and quality at every step of the processing or manufacturing journey, from the receipt of raw materials (such as primary products and other ingredients) to the shipping and marketing of the final consumer products.

Implementing GHPs involves the use of appropriate sanitary measures designed to prevent microbial contamination, ensuring optimal sanitary conditions during the food processing stages. Key aspects of GHPs include:

- The application of effective cleaning and sanitizing methods, using approved agents at the correct concentration and frequency to prevent microbial buildup on processing equipment, utensils, or food contact surfaces.
- Strict observation of sanitary practices, proper use of protective clothing, and adherence to personal hygiene standards by all personnel involved in handling and processing food.
- Providing hand-washing and hand-sanitizing stations when necessary.
- Implementing time and temperature controls to inhibit microbial growth in susceptible intermediate and finished products.
- Employing other specific sanitary measures needed due to the nature of the food being processed, the technology used, or the characteristics of the facilities where processing takes place.

GMPs incorporate practices that ensure:

- All food materials and ingredients, including food additives, meet quality and safety standards before use and are stored correctly to avoid contamination or mix-ups.
- Food production facilities are the right size to prevent overcrowding, allowing proper storage and placement of equipment, raw materials, and other product components like packaging and labeling.
- Facility layouts support the efficient flow of production materials and personnel.
- Adequate lighting is provided in processing areas.
- Equipment is properly maintained to ensure it functions correctly.
- Essential parameters like temperatures, processing times, pressures, and machine operations are controlled according to specifications to ensure proper processing.
- Accurate labels are applied.

These control procedures also involve examining or sampling intermediate foods during processing and finished products from final storage. These products are tested or analyzed to ensure they meet product specifications, as well as safety and quality standards.

When applied correctly, GMPs also include the establishment of record-keeping systems to track the outcomes of quality control activities. Recorded data may include:

- Results from quality assurance inspections of production facilities before and during production.
- Processing parameters, such as cooking times, temperature records, and pressure levels.
- Results from on-line product inspections (e.g., net weights, can seal checks).
- Examination results regarding the integrity of package closure systems.
- Specific laboratory testing methods for determining product quality and safety, including sample size and acceptance/rejection criteria for each batch.

Some food processing techniques are highly complex, while others are more straightforward. Each process must be carefully evaluated for potential foodborne hazards and the potential impact on food quality and safety if a processing failure occurs. Such failures could result in unacceptable risks to consumer health.

PROCEDURES SAFETY CONTROL

In some cases, because of the nature of the food processing methods or the hazards associated with some foods, consideration is given to applying intensified safety control procedures or systems. One such system is that based on the Hazard Analysis and Critical Control Point (HACCP). In order for the HACCP system to be effective, there must first be an effective GHP and GMP system in place.

HACCP includes the identification of all the known potential hazards which can be associated with the food being processed. Once this hazard assessment is done, critical control points (CCPs) in the processing are identified where controls can be exercised to prevent, reduce or eliminate these hazards. Constant vigilance is maintained over the CCPs to prevent any process deviations that would result in loss of control at the CCP. Appropriate corrective actions are required whenever a CCP is found to be out of control, and the suspect food product is generally prevented from being distributed until its safety and acceptability have been determined. This system is highly effective when employed properly, but it requires considerable understanding and technical information related to the food product, the processing methods and the production facility.

In manufacturing and processing of foods it is also necessary to ensure the safety of ingredients used as technical aids, additives, flavourings or colourings. Such safety assessments require the analysis of test data, chemical specifications for substances involved and information on human dietary consumption levels and patterns. It is also necessary to evaluate the impact of uncertainties in cases where the information is insufficient to make a clear safety assessment decision. This is the work of highly trained specialists in toxicology, nutrition, chemistry, food composition and risk assessment techniques. The required expertise is often found only in countries with highly trained personnel and advanced technological capabilities.

Technical Assistance Needs

Food technology and science are complex, involving specialized knowledge in a wide range of fields such as chemistry, biochemistry, physical chemistry, microbiology, nutrition, toxicology, physics, radiology, statistics and mathematics. As a consequence, food control measures are diverse and complicated. The technical dimensions are different for nearly every food product, for the various technologies used in food preparation, processing and manufacturing and for the various types of facilities in which food is produced.

In view of the many food safety concerns of consumers and the diversity in scope and dimensions of food quality and safety problems, technical assistance is often needed. Furthermore, new food products are created every day and new technologies are being developed and introduced rapidly, so the demand for keeping up with the scientific advancements in food technology is high. Emerging hazards such as antibiotic-resistant microbes and novel pathogenic bacteria present food control officials with new challenges in maintaining controls to ensure public health. With rapid shipping methods and the global distribution of food, serious public health risks and food hazards in one part of the world can be transferred to other parts of the world in a matter of hours or a few days.

Consumers expect government to look after their interests in making sure that the food industry produces safe food and that economic fraud, unfair trade practices and risks to human health are minimized. Government frequently does not have the financial and technical resources to provide such assurance, especially in developing countries. Many developing countries lack access to the latest knowledge and information about new food processing technologies. They may also lack technically trained staff, equipment, methods and facilities for testing or analysing food for contaminants, toxins, chemical or drug residues or microbiological contamination.

In some countries, there is a need to update and revise the existing legal framework regarding food quality and safety. Regulations governing food standards are often lacking or outdated. Food control infrastructure may be non-existent, poorly organized or inadequately supported because of the lack of sufficient financial resources. In many countries, different government ministries or agencies are involved in food regulation and control, but their failure to coordinate their activities results in a waste of resources because of overlapping and redundant work efforts. There is generally a need for improved regulatory food inspection and laboratory services, development of food control enforcement programmes and the administration and coordination of food control activities in developing countries. Training in technical areas of food control is nearly always needed in such countries.

QMS Frameworks

Several frameworks exist to guide QMS development, including:

- ISO 9001
- ISO 9000 series
- ISO 14000 series (environmental management systems)
- ISO 13485 (quality management systems for medical devices)
- ISO 19011 (auditing management systems)

- IATF 16949 (quality management systems for automotive-related products)
- AS 9100 (quality management systems for aviation, space, and defense)

Several industries also have developed standards, built upon ISO 9001, that offer specific direction regarding extra expectations that are particularly associated with their operations.

ISO 9001 is the most recognized and implemented quality management system standard in the world. ISO 9001 specifies the requirements for a QMS that organizations can use to develop their own programs.

The standard is structured around seven quality management principles. Quality management principles are the guiding beliefs that, when used for organizational decision-making, ensure long-term success from the perspective of customers, employees, and other stakeholders. The quality management principles demonstrate the depth of the organization's commitment to quality. Well-formulated principles should be timeless, not dependent on temporary circumstances.

The principles of ISO 9001:2015 are as follows:

- Customer focus
- Leadership
- Engagement of people
- Process approach
- Continuous improvement
- Evidence-based decision making
- Relationship management

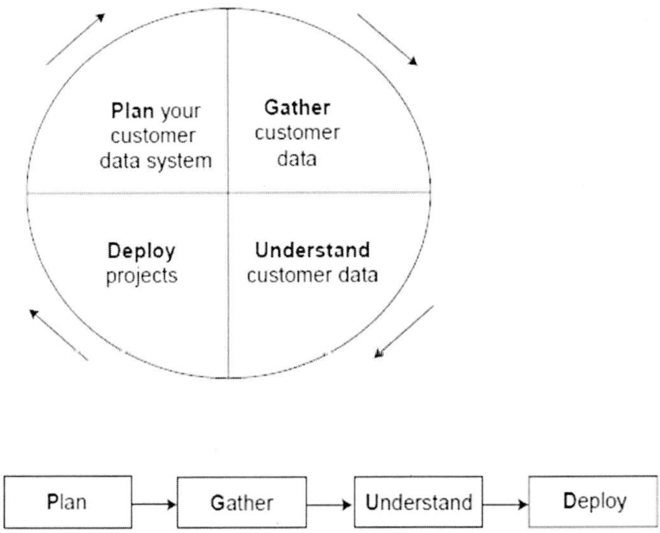

Figure 11.4 A Customer Satisfaction Measurement Process Shown as a Continuous Cycle

The Components and Prerequisites of a QMS

Each element of a QMS helps achieve the overall goal of meeting customer and organizational requirements. A QMS should address an organization's unique needs, but many of the QMS systems have common expectations, including:

- Leadership
- Planning, commitment, and review
- The system spans the breadth of the organization
- Metrics and feedback
- Communication, education, and training
- Risk management
- Teamwork
- Compliance to requirements
- Continuous improvement

How is a QMS Establish Implement

Every organization has its own culture, which influences which QMS methods are most efficient and effective. Instead of developing a QMS by addressing each expectation described in a selected model, organizations should outline the approaches currently in use for managing quality and meeting customer expectations.

A basic model for gathering customer expectations can be structured in the following four steps: plan your customer data system, gather customer data, understand the data, and deploy (make use of) the data.

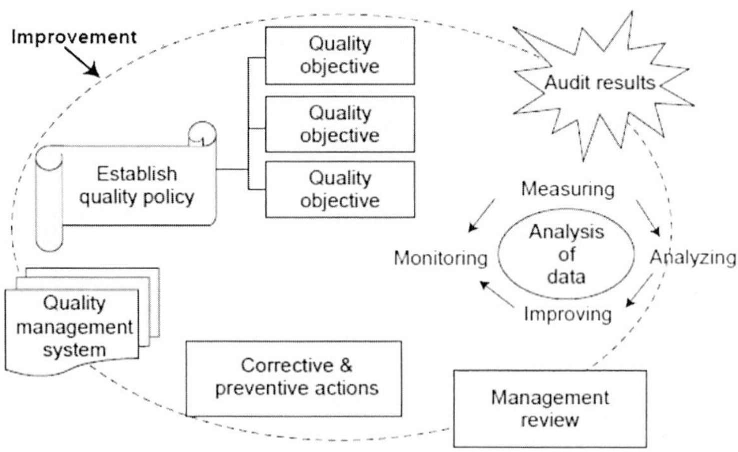

Figure 11.5 The improvement loop

Next, map those approaches against the appropriate QMS framework or model. For each identified gap, the organization should determine whether it would benefit from addressing the opportunity, and must clearly understand the potential benefit before implementing any changes.

The QMS design should be influenced by the organization's objectives, needs, products, and services. This structure is based largely on the plan-do-check-act cycle and allows for continuous improvement to the product and QMS.

The basic steps to implementing a QMS are as follows:

- Design
- Build
- Deploy
- Control
- Measure
- Review
- Improve

Design and Build

The design and build steps develop the QMS's structure, processes, and implementation plans. Top management should oversee these steps to ensure the needs of the organization and its customers are a driving force behind the system's development.

Deploy

Deployment is best served granularly by breaking down each process into subprocesses and educating staff on documentation, education, training tools, and metrics. Organizational intranets are increasingly being used to help deploy QMSs.

Control and Measure

The realization of these two areas is accomplished largely through implementing routine, systematic audits of the QMS. The specifics vary from organization to organization, depending on size, potential risk, and environmental impact.

Review and Improve

These steps detail how the audit results are handled. The goals are to determine the effectiveness and efficiency of each process toward its objectives, communicate these findings to the employees, and develop the best new practices and processes based on the data collected during the audits.

The improvement loop, shown below, indicates how a QMS provides the necessary pathways for organizations to implement quality objectives via quality policies and procedures. Opportunities for improvement are identified via quality audits, and applicable corrective and preventive actions are developed and implemented. Proper data analysis helps steer improvement efforts. The overall effectiveness of the improvement loop is discussed in management reviews.

Safety Control Procedures

In some cases, because of the nature of the food processing methods or the hazards associated with some foods, consideration is given to applying intensified safety control procedures or systems. One such system is that based on the Hazard Analysis and Critical Control Point (HACCP). In order for the HACCP system to be effective, there must first be an effective GHP and GMP system in place.

HACCP includes the identification of all the known potential hazards which can be associated with the food being processed. Once this hazard assessment is done, critical control points (CCPs) in the processing are identified where controls can be exercised to prevent, reduce or eliminate these hazards. Constant vigilance is maintained over the CCPs to prevent any process deviations that would result in loss of control at the CCP. Appropriate corrective actions are required whenever a CCP is found to be out of control, and the suspect food prod

TRACKING THE FOOD SUPPLY CHAIN

Food supply chain tracking, also known as food traceability, is the process of monitoring the movement of food products from production to consumption. It involves recording and linking the steps of production, processing, and distribution.

Food traceability is the ability to follow the movement of a food product and its ingredients through all steps in the supply chain, both backward and forward. Traceability involves documenting and linking the production, processing, and distribution chain of food products and ingredients. In the case of a foodborne illness outbreak or contamination event, efficient product tracing helps government agencies and those who produce and sell food to rapidly find the source of the product and where contamination may have occurred. This enables faster removal of the affected product from the marketplace, reducing incidences of foodborne illnesses.

Figure 11.6 Food supply chain tracking

Consider that by scanning the unique code on each food product, you can access information including the product batch number, the production and expiry dates, the region that the raw material was sourced from and a certificate of inspection. Being able to determine where food has come from and that it is safe to consume is important to an end consumer. There is also a strong link between authenticity and producc origins, and food safety and quality.

Traceability is a paradigm shift in modern food supply chain. And the technology is already available to trace each raw material that goes into a product and follow how a product is used and where it is discarded. The rise of the internet of things (IoT) will also help in tracking of food products in supply chain. Sensors are coming onto the market that track not only the location of goods, but also the orientation of crates (containing food packets) and factors such as temperature and humidity. Traceability enables companies to meet and balance a broader set of business objectives, including efficiency, resilience, responsiveness, and sustainability.

Traceability gives companies the ability to follow products and goods as they move along the value chain and to extract information about the origin of inputs, supplier sourcing practices, and manufacturing processes. With that data in hand, companies can make predictions, run production scenarios, and streamline operations. Companies can also identify strategic supply chain opportunities, innovate faster, minimize the impact of internal and external disruptions, and certify sustainable processes and products.

Food Sector – Supply Chain in India

The food and agriculture sector are the single largest employer in India. India witnesses nearly 5-15 % (About $ 13 Billion in value) wastage in fruits and vegetables annually, due to lack of modern harvesting technologies and cold chain infrastructure. And a rising population, expected to reach almost 1.6 billion by 2030, exacerbates our already strained food systems. Hence, it is very important to focus on the supply chain to figure out the inefficiencies and weed them out so that our food supply chain is able to withstand the growing demand.

People ask about relationship between sustainability and traceability, and it is often misunderstood. Traceability is the ability to follow materials from the beginning of the supply chain to the customer who purchases a product. With this data and information, traceability plays two roles that support sustainability. It provides visibility on inputs and processes across the value chain, and it provides the source information for its origins and sustainability certifications. It could potentially facilitate comprehensive tracking of the environmental, economic, health and social consequences of different agricultural production processes, even making it possible to calculate the true cost of food to consumer, which will help meet consumer demand for transparency. Consumers might favour a locally grown produce than a produce, which has travelled a long distance to the retail shelf. In addition, producers, including small-scale producers, also look to harness potential efficiencies brought about through traceability-enabled transparency, such as associated cost savings and new value sources. Traceability could also improve farmers/producer revenue, market access and opportunities for affordable access to capital.

Initiatives of Government of India

Traceability provides a significant opportunity to create value for consumers, producers, and supply-chain operators. One key opportunity is the potential to identify and address loss points in the supply chain, which is a particular challenge for India, and it holds a distinct promise in helping drive food-systems transformation. It has the potential to strengthen inclusivity and empower small-scale producers with improved market visibility and access to new services and resources. However, to ensure that its potential impact is maximized, the traceability agenda must focus on ways to introduce inclusive scaling, particularly in relation to medium and small-scale industries.

With this in mind, Ministry of Food Processing Industries has launched Operation Greens with the following objectives:

- Enhancing value realisation of farmers by targeted interventions to strengthen production clusters and their Farmer Producer Organisations and linking/ connecting them with the market.
- Price stabilisation for producers and consumers by proper production planning in the clusters.
- Reduction in post-harvest losses by creation of farm gate infrastructure, development of suitable agro-logistics, creation of appropriate storage capacity linking consumption centres.
- Increase in food processing capacities and value addition in value chain with firm linkages with production clusters.
- Setting up of a market intelligence network to collect and collate real time data on demand and supply and price of crops.

The scheme will have two-pronged strategy of price stabilisation measures (for short term) and integrated value chain development projects (for long term).

Circularity: A Growing Industry Focus

Circularity is a recent industry trend that broadly encompasses the reuse, restoration, and recycling of industrial waste. It has become a critical topic of discussion among industries worldwide as companies seek sustainable waste management solutions.

During a discussion with a global food processing company planning to establish a plant in India, they expressed concerns about wastewater management. Recognizing future waste management trends, they are making a significant investment in a water recycling facility. Similarly, an Indian FMCG conglomerate is addressing the challenge of plastic waste generated from its packaging. Both companies are actively exploring sustainable strategies to minimize environmental impact.

Capturing the economic and material value of industrial waste, rather than allowing it to become an environmental burden, requires a systemic approach that optimizes opportunities across the food supply chain. Effective waste tracking and management solutions can unlock new investment prospects in the manufacturing sector.

Implementing Food Traceability Pressnts Challnge

Full-scale traceability would be difficult for a food value chain player to achieve alone, as barriers to adoption are significant on a large scale (the financing of capital expenditures and establishment of consistent standards, usage of technology). Collaborations will likely come in one of three forms: horizontal (the largest global retailers and food companies work together to align on standards and put pressure on the rest of the supply chain to comply), vertical (one retailer collaborates with their full supply chain to align on standards) or multi-stakeholder (companies collaborate with players from throughout the entire value chain with support from government and consumers). Multi-stakeholder collaboration is preferable to the other two forms because it helps set consistent standards, taking into consideration the needs and perspectives of the full supply chain.

Traceability in food value chains, coupled with multi-stakeholder collaboration, has the potential to fundamentally improve food supply chains. To achieve its full potential, stakeholders will need to come together to enable emerging technologies and to install a broad system and standards in traceability. Without multi-stakeholder collaboration, the impact of traceability could be limited. Traceability requirements could become operationally and financially burdensome, with complex and inconsistent standards, providing limited benefit to consumers and stakeholders alike. For example, if downstream players pursue traceability in silos, food producers will need to comply with several different types of requirements and the associated costs. This risk is already apparent in the proliferation of voluntary market standards for different foods (e.g., organic food standards, Food standards/certifications for Import/Export) and if horizontal collaboration between retailers is not deliberately designed to be inclusive, it could exclude small-scale producers from markets entirely.

In the last few years, the Agricultural and Processed Food Products Export Development Authority (APEDA) has been shown growing interest in electronic traceability system for farming products while simultaneously providing guidelines to the farmers about good agricultural practices as well as focusing over the food safety standards for improving better production and distribution. Nowadays, building up traceability system for exports has become easier with help of APEDA Tracenet Module. These steps will go a long way to implement traceability across food supply chains in India.

India has come a long way in the journey of food traceability in the last decade with initiatives from private as well as public sector. There is still a long journey ahead for food traceability India.

Once implemented across supply chains, food traceability will impact our supply chains in the following ways:

- Meet consumer demand for food production transparency.
- Enhance the ability to identify, respond to and even prevent food safety issues.
- Support supply-chain optimization and reduce food loss.
- Validate sourcing claims to support sustainability goals.

However, the most important questions today are – which food company will take the lead in implementing traceability in food supply chain? Is the consumer ready to bear the higher cost owing to increased usage of technology in implementing traceability? What will be the role of government regulators in implementing traceability in food supply chain? There are no easy answers to these questions, however, we as a consumer must ask all the stakeholders to implement traceability in food supply chain. No single company will be able to solve the complex issues underpinning traceability agenda alone. Building sustainable value chains in traceability will take a strong commitment and collaboration between public and private players.

For companies at fence evaluating their options, it's time to act. For those piloting and experimenting, it's time to scale. And for companies operating full-scale projects, it's time to scale up the benefits and share the learnings with other industry players. It is also time for Indian food processing sector to be at the forefront of the upcoming food supply chain revolution!

CHALLENGES FACING THE FOOD SUPPLY CHAIN IN INDIA

India's food supply chain encounters multiple obstacles that affect its overall efficiency and reliability. These challenges stem from inadequate infrastructure, climate change, food wastage, and regulatory constraints, all of which influence how food is produced, transported, and consumed:

- **Infrastructure Deficiencies:** One of the most pressing issues is the lack of proper infrastructure, particularly in rural areas where road networks are poorly developed. Inadequate transportation facilities make it difficult to move food from farms to markets in a timely manner. This is especially problematic for perishable items like fruits, vegetables, and dairy, as long transit times often lead to spoilage before the products reach urban centers.
- **Cold Storage Limitations:** The shortage of refrigerated storage and transportation facilities exacerbates food losses. While India is one of the world's largest producers of fresh produce, a significant portion goes to waste due to the absence of an efficient cold chain system. Post-harvest losses not only diminish farmers' earnings but also contribute to higher food prices. Expanding and upgrading cold storage units, along with improving road connectivity and logistics, would significantly enhance the efficiency of the supply chain.
- **High Levels of Food Wastage:** India faces a substantial food wastage problem, leading to economic losses and worsening food insecurity. Estimates suggest that nearly 40% of the food produced is wasted due to inefficiencies in storage, transportation, and supply chain management. Perishable commodities such as fruits and vegetables are particularly vulnerable, requiring proper handling and refrigeration to maintain freshness.
- **Wastage at Multiple Stages:** Food losses occur at various points in the supply chain—during harvesting due to outdated techniques, in transit due to inefficient logistics, and at markets due to oversupply or lack of demand. Addressing these inefficiencies would not only reduce wastage but also help

farmers secure better prices for their produce, ensuring that a larger proportion of food reaches consumers in good condition.

- **Regulatory Hurdles:** India's food supply chain must comply with numerous food safety and quality regulations, which, while necessary, can be challenging for small-scale farmers and businesses. Many small producers lack the financial resources and technical know-how to meet stringent food safety laws, obtain quality certifications, and adhere to export regulations.
- **Compliance Challenges:** Standards such as those set by the Food Safety and Standards Authority of India (FSSAI) ensure food safety, but meeting these requirements involves costs related to testing, certification, and process modifications. Small and medium enterprises (SMEs) often struggle with bureaucratic hurdles and complex paperwork, making compliance a difficult and costly process.
- **Impact of Climate Change:** Climate change poses a significant threat to India's food supply chain, as agriculture in the country heavily depends on monsoon rains. Any disruption in rainfall patterns can have devastating effects on food production. Unpredictable weather events such as floods, droughts, and erratic rainfall can drastically reduce crop yields, leading to supply shortages.
- **Effects on Agriculture:** Rising temperatures also negatively impact crops and livestock, reducing productivity levels. Farmers are increasingly witnessing shifts in growing seasons and changes in crop patterns, requiring them to adopt new farming techniques. However, many small farmers lack the financial means and technological access to implement climate-resilient agricultural practices, making them particularly susceptible to the adverse effects of climate change.

Addressing these challenges requires a multi-faceted approach involving infrastructure improvements, better cold storage facilities, regulatory support for small farmers, and climate adaptation strategies to ensure a more resilient and efficient food supply chain in India.

Government Initiatives and Their Impact on the Food Supply Chain

The Indian government has launched multiple initiatives to enhance the efficiency, sustainability, and inclusivity of the nation's agriculture and food supply chain. These efforts aim to support farmers, improve market accessibility, and incorporate technology to make agriculture more productive and resilient. Two major programs, PM-KISAN and eNAM (National Agriculture Market), have significantly contributed to transforming the food supply chain in India.

The Pradhan Mantri Kisan Samman Nidhi (PM-KISAN) is a flagship scheme introduced to provide direct financial assistance to small and marginal farmers. Under this program, eligible farmers receive ◎6,000 annually in three equal installments. By addressing the financial constraints faced by farmers with limited resources, this initiative ensures they have access to essential agricultural inputs such as seeds, fertilizers, and equipment. The direct income support plays a crucial role in stabilizing farm operations and improving livelihoods.

eNAM, or the National Agriculture Market, is a digital platform that connects various mandis (agricultural markets) across India. By integrating these markets into a unified network, eNAM enables farmers to sell their produce directly to buyers across the country, eliminating intermediaries and improving price transparency. This initiative enhances farmers' bargaining power, ensuring they receive fair market rates and greater access to diverse markets.

Beyond government programs, public-private partnerships (PPPs) have emerged as a key factor in strengthening the food supply chain. By leveraging the expertise, technology, and resources of both the public and private sectors, PPPs drive innovation and efficiency in agricultural practices, supply chain logistics, and market accessibility. These collaborations contribute to long-term sustainability and improved agricultural productivity, ensuring that India's food supply chain becomes more resilient and future-ready.

The Necessity of Agriculture Traceability

The supply chain for a product can be so intricate that traceability is a prerequisite for many industries. Due to traceability, suppliers are able to respond swiftly to any challenges by maintaining a record of the full production and distribution history. For instance, in the event of a product recall, suppliers can identify the cause of the issue and direct distributors to take the product off the market. This shields both the consumer and the provider from any legal action and potentially hazardous products.

Additionally, a product's traceability can be utilized to demonstrate particular characteristics, such as its seeding procedures, country of origin, storage history, and much more. Transparency is desired when consumers become more knowledgeable about how goods are manufactured and the problems related to manufacturing.

The Role of Traceability in the Food Supply Chain

Traceability requirements for the food supply chain are essential for helping businesses stay competitive in both domestic and foreign markets while taking public health into consideration. The ability to track a product through every stage of production—from distribution and transit to retail and ultimately to the customer—is essential for any business that deals with produce.

Food traceability systems essentially cover the entire production process, from farm to table. Since upholding global traceability standards is just as difficult as upholding product traceability, effective traceability solutions are essential in the food industry.

Traceability Standards in India: Ensuring Transparency and Accountability

Traceability standards in India are frameworks designed to ensure the complete transparency of the origin, movement, and transformation of products through the supply chain. These standards are crucial in sectors like food safety, pharmaceuticals, agriculture, and manufacturing to provide accountability, improve quality control, and protect consumer health. With rapid industrialization and globalization, traceability systems have gained importance in maintaining the integrity of products and enhancing safety.

In India, the concept of traceability is regulated and enforced by various government agencies and organizations depending on the industry. For example, the Food Safety and Standards Authority of India (FSSAI) plays a key role in ensuring food traceability. The FSSAI mandates food business operators to maintain records that trace the origin of food ingredients, production processes, and distribution channels. This is crucial for managing food recalls and ensuring consumer safety, especially in the event of contamination or fraud.

Similarly, in pharmaceuticals, the Central Drugs Standard Control Organization (CDSCO) regulates traceability through the use of technologies such as barcodes and RFID (Radio Frequency Identification). These systems help track the movement of drugs from manufacturers to consumers, reducing counterfeit medicines and ensuring that all products comply with safety standards.

The agriculture sector also has its own set of traceability guidelines. The Agricultural and Processed Food Products Export Development Authority (APEDA) oversees traceability in the export of agricultural products to ensure that they meet international standards. This includes ensuring that all agricultural exports are traceable from farm to the point of sale, with documentation to prove compliance with health and quality standards.

Traceability systems in India often rely on technology like blockchain, RFID, and GPS tracking to enhance transparency and accountability across the supply chain. Blockchain, in particular, is gaining traction for its ability to provide a secure, immutable record of transactions that can be accessed by all stakeholders in the supply chain, ensuring a high level of trust.

Overall, traceability standards in India are becoming increasingly important as industries seek to enhance product quality, consumer safety, and regulatory compliance. These standards promote greater transparency, reduce risks associated with fraud and contamination, and improve consumer confidence in domestic and international markets.

Traceability's Function in the Food Supply Chain

The need for traceability in the food supply chain is essential for companies to stay competitive in both domestic and international markets while prioritizing public health. Any company handling produce must be able to trace a product through all stages of production, including distribution, transit, retail, and ultimately to the consumer.

Essentially, food traceability systems cover every step of the process, from farm to fork. Effective traceability solutions are vital in the food industry, as ensuring compliance with global traceability standards is just as complex as maintaining the traceability of individual products.

What Elements Make Up The Food Traceability System?

Traceability's two primary elements are Tracking and Tracing. Tracking refers to the location of a certain product unit or batch at a specific point in the production process. It starts at the processing stage and continues through distribution and end-user consumption in a food supply chain, also known as the downstream path.

Tracing, often known as the upstream path, is the documentation of a product's journey from its point of origin to the ultimate consumer.

How can Traceability help Farmers?

Traceability has many known advantages. However, they might act as trade barriers for small-scale farmers in developing nations, particularly those that grow horticulture and other fresh goods. Small-scale agricultural producers who lack the resources to adhere to increasingly stringent standards, especially those requiring the tracking and monitoring of environmental and supply chain variables through sophisticated technologies, may be excluded from the market for safe and traceable food.

For farmers, traceability is a useful commercial strategy. Product traceability is a requirement, not an option and the key is to figure out the best way to accomplish this and seize the possibilities that are present.

Solutions for Food Traceability in the Future

Both humans and the planet rely on food for survival. To address the complex challenges within the food system and build a sustainable, healthy, and nutrient-rich food ecosystem, a groundbreaking approach is necessary. Technology plays a pivotal role in accelerating this transformation. Blockchain-enabled food traceability solutions provide a solid foundation to address these challenges and revolutionize the agri-food industry. By ensuring transparency in food production, traceability helps monitor environmental, social, and economic factors throughout agricultural value chains to meet the growing demand for clear, traceable food.

Key benefits of food traceability include the following:

- Enhancing transparency with customers.
- Identifying and resolving food safety issues.
- Streamlining the supply chain and reducing losses.
- Validating claims about sustainable practices.
- Transforming food supply chains through technology-driven traceability systems that continuously gather data.
- Providing a secure, distributed ledger that allows for data sharing and analysis.
- Fostering healthy, transparent collaboration among stakeholders, manufacturers, and customers.

Bibliography

Agarwal, R., & Joshi, M. (2018). Efficiency of Warehouse Management Systems in Indian Agriculture. Indian Journal of Warehousing and Storage, 32(4), 201-208.

ASSOCHAM. (2020). Role of Logistics in Indian Agricultural Supply Chains. Associated Chambers of Commerce and Industry of India (ASSOCHAM).

ASSOCHAM. (2021). Post-Harvest Supply Chain and Logistics Solutions for Indian Agriculture. Associated Chambers of Commerce and Industry of India (ASSOCHAM).

Confederation of Indian Industry (CII). (2021). Logistics and Post-Harvest Solutions in Indian Agriculture. CII Report.

Confederation of Indian Industry (CII). (2021). Post-Harvest Technology and Logistics in Indian Agriculture. CII Report.

Directorate of Marketing and Inspection, Government of India. (2020). Post-Harvest Infrastructure: Warehouse Management and Logistics in Indian Agriculture. Government of India Report.

Directorate of Marketing and Inspection, Ministry of Agriculture. (2020). Improvement in Post-Harvest Infrastructure and Logistics. Government of India.

FICCI. (2020). Supply Chain Solutions for Indian Agricultural Products. Federation of Indian Chambers of Commerce and Industry.

Food and Agriculture Organization of the United Nations. (2020). Global Post-Harvest Management: Indian Case Studies. FAO Report.

Food Corporation of India (FCI). (2021). Logistics in Agricultural Supply Chains: Role of Warehouses and Transportation. FCI Publications.

Food Corporation of India. (2021). Agri-Warehousing and Post-Harvest Logistics in India: A Strategic Approach. FCI Publications.

Food Processing and Agro-Logistics Association. (2019). Indian Agricultural Logistics: Post-Harvest and Beyond. FAAL Reports.

Government of India. (2019). Food Security and Logistics Management Act. Ministry of Food Processing Industries.

Government of India. (2019). Warehouse Development and Regulatory Authority Act: Post-Harvest Infrastructure. Ministry of Consumer Affairs.

Government of India. (2020). National Policy on Post-Harvest Management. Ministry of Agriculture and Farmers Welfare.

Government of India. (2021). National Agricultural Market Policy and Infrastructure Development. Ministry of Agriculture.

Government of India. (2022). National Agricultural Market Development and Logistics Infrastructure. Ministry of Agriculture and Farmers Welfare.

Gupta, S., & Mehta, R. (2021). Logistics Optimization for Post-Harvest Agriculture in India: A Critical Analysis. Journal of Indian Agricultural Engineering, 28(5), 128-135.

Gupta, V., & Mehta, H. (2020). Agricultural Supply Chain Management: Addressing Post-Harvest Losses in India. International Journal of Food and Agricultural Logistics, 24(5), 51-59.

ICAR. (2021). Best Practices for Post-Harvest Management in Indian Agriculture. Retrieved from http://www.icar.org.in/post-harvest-management

Indian Cold Chain and Logistics Association. (2021). Cold Chain Solutions for Post-Harvest Food Security in India. Retrieved from http://www.iccla.org/cold-chain-solutions

Indian Council of Agricultural Research (ICAR). (2019). Advances in Post-Harvest Management of Fruits and Vegetables. ICAR Publication.

Indian Council of Agricultural Research (ICAR). (2021). Post-Harvest Processing and Storage: Policies and Guidelines for India. ICAR Publication.

Indian Institute of Management Ahmedabad. (2020). Agricultural Supply Chains in India: Opportunities and Challenges. Retrieved from http://www.iimahd.ac.in/agri-supply-chain

Indian Institute of Management Ahmedabad. (2020). Supply Chain Strategies for Post-Harvest Handling in India. Retrieved from http://www.iima.ac.in/agri-supply-chain

Jadhav, M. (2021). Managing Post-Harvest Losses in Indian Agriculture. Food Tech Books.

Jain, R. (2016). Logistics and Transportation in Post-Harvest Agriculture: An Indian Perspective. Research & Development Publications.

Jain, R., & Kumar, P. (2019). Logistics and Supply Chain Integration for Agricultural Products. Proceedings of the International Conference on Agri-Logistics, Hyderabad, 121-125.

Kader, A. A. (2005). Increasing Food Availability by Reducing Postharvest Losses of Fresh Produce. Acta Horticulturae.

Kitinoja, L., & Kader, A. A. (2002). Small-Scale Postharvest Handling Practices: A Manual for Horticultural Crops (4th ed.). University of California, Davis.

Ministry of Food Processing Industries (MoFPI). (2020). Mega Food Park Scheme Guidelines. Government of India, New Delhi.

Nanda, S. K., & Kumar, R. (2018). Agri-logistics in India: Challenges and Scope. Indian Journal of Agricultural Economics, 73(3), 320–329.

Narayan, S., & Rajendran, P. G. (2019). Post-Harvest Technology and Value Addition of Horticultural Crops. New India Publishing Agency, New Delhi.

Singh, R. P., & Heldman, D. R. (2014). Introduction to Food Engineering (5th ed.). Academic Press.

Sinha, A., & Roy, S. (2016). Post-Harvest Management in Horticultural Crops: Technologies and Strategies. Agrotech Publishing Academy, Udaipur.

Tiwari, G. N., & Jain, D. (2007). Drying Technologies for Foods: Fundamentals and Applications. Narosa Publishing House.

Verma, L. R., & Joshi, V. K. (2000). Postharvest Technology of Fruits and Vegetables: Handling, Processing, Fermentation and Waste Management. Indus Publishing.

Index

V

W